For the Warning of the Earth

For the Warming of the Earth

Music, Faith and Ecological Crisis

Mark Porter

scm press

© Mark Porter 2024
Published in 2024 by SCM Press

Editorial office
3rd Floor, Invicta House,
110 Golden Lane,
London EC1Y 0TG, UK

www.scmpress.co.uk

SCM Press is an imprint of Hymns Ancient & Modern Ltd
(a registered charity)

HYMNS Ancient
&Modern

Hymns Ancient & Modern® is a registered trademark of
Hymns Ancient & Modern Ltd
13A Hellesdon Park Road, Norwich,
Norfolk NR6 5DR, UK

British Library Cataloguing in Publication data
A catalogue record for this book is available
from the British Library

ISBN 978-0-334-06568-5

Typeset by Regent Typesetting
Printed and bound by
CPI Group (UK) Ltd

Contents

Acknowledgements

This book is the result of numerous conversations, and very few of the ideas that fill its pages come from me alone. Thanks must go first of all to everyone who made the time to talk to me. It is their creativity that inspired me to explore deeper into these themes, and it is their work that has allowed me to put together this collection of voices and perspectives. I am grateful to all those I interviewed for their willingness to share their stories and experiences with someone many of them had only just met.

Much of the work behind this book was unsupported by any formal academic institutions. In their place I must, once again, thank my parents for a place to live and all kinds of support above and beyond. Thanks are also due to members of the Refo community in Berlin, who hospitably offered me a room just when I needed it, and to Monique Ingalls and Lester Ruth for their generous support through Patreon. Anika Mund has been there for me throughout much of the writing process, she has inspired me with her love for animals, and has always trusted that this work was something worth writing. The many individuals and projects who help to provide access to academic materials for those without institutional access to them also deserve some recognition, whether or not they can individually be named in print.

I want to thank the students on my ecomusicology courses in Cologne and Erfurt for the chance to discuss a range of perspectives on music and ecology on a weekly basis. Discussing these themes together helped to clarify my thoughts on a whole range of topics and gave me the chance to explore themes I would never otherwise have encountered. Stella Lorenz played an important role in making the course in Erfurt both possible and enjoyable, and my thanks go to her for helping to make that happen. I am grateful to the individuals who offered me freelance work over the course of writing that helped to keep me fed and warm. Thanks also go to Patrick Becker at the University of Erfurt for believing in my work enough to offer me a job. The relative stability of a fixed-term contract has given me the chance to focus on the later stages of this project without the continual worry of where my next piece of work is going to be coming from.

Finally, it seems strange to me for a book focused on ecology to owe thanks only to members of the human species. It is important, therefore, to acknowledge the ways in which encounters with landscapes, with plants, with insects and with other animals have contributed to my own engagement in this area. In particular, I take inspiration from the ecosystems around Erfurt in East Germany, Cumnor in Oxford and Brighstone on the Isle of Wight. Thank you to all who inhabit these environments – my encounters with you are always important and have helped me on the way to wanting to research something new. This Earth and these lands are your home as much as they are mine, and I hope to do increasing justice to that reality.

Introduction

A Surprising Combination of Interests

In the autumn of 2019 I was looking for a job. I had come to the end of a series of short-term post-doctoral contracts, and I wasn't really sure what I might do next. Idly scrolling through Facebook and Twitter as I put together job applications and wondered what the coming months might bring, a number of events and projects from different Christian groups started to cross my social media feeds. While I was not particularly surprised at an uptick in projects relating to climate change or the natural world, I *was* a little surprised by the proportion of these that seemed to bring together climate change and music. Christian Climate Action featured music prominently in several events over the course of 2019; a number of different requiems for lost species had been performed or were in processes of commissioning; and the evangelical song-writing collective Resound Worship put out a call for new songs centred on creation.

While one or two small events might have been easy to ignore, the wide range of different projects drew me in and, as someone with a long-standing interest in music and Christianity, I was naturally intrigued. The natural world and issues of climate change have for me, as for many others, become increasingly important in recent years, and the chance to bring this interest together with my existing expertise seemed to deserve some kind of exploration. Without really knowing where any of this might lead, and without any particular intention to begin another multi-year research project when it seemed that the university system had, at least for the moment, given up on funding me, I sent out a couple of emails to some of the people involved in the different projects. In the face of a climate crisis that is far from musical in origin, I was intrigued as to why so many people were now turning to music, and I wanted to know how exactly they went about creating something that was fitting for the current ecological moment. What were they thinking about when they put pen to paper? What different priorities shaped the music they were writing? Did they believe that Christian music and climate change really had any hope of coming together in a meaningful manner? I had no real mindset for thinking about these questions – my scholarship up to this point had not prepared me with ready answers – and I was genuinely intrigued as to what they would tell me.

By the end of 2022 I had engaged in more than 40 conversations around a range of different musical projects both in the UK and farther afield. I had spoken to activists who had been imprisoned for their protests, to songwriters grappling with a new realm of creativity and to composers commissioned for high-profile occasions. I had heard from amateurs unsure of their musical abilities and from those avoiding music in order to better hear the natural world around them; from those pragmatically focused on motivating change and from those unsure whether change is even possible. While, at the start, I may have doubted how much there was to say or whether these themes were really worthy of investigation, by the end it seemed there were so many different things to think about that it can be hard to know even where to begin.

Through the course of the conversations, I began to see how music can be brought to bear on the climate crisis in interesting ways and how, through music, we can gain new perspectives on some of the key issues we are all seeking to navigate. Through music, we enter a microcosm for broader issues surrounding ritual, liturgy, worship and our relationship to the natural world. Through singing, listening and sounding on our own or together with others we enter into an intersectional realm that draws together questions of feeling, community, spirituality, ethics and much, much more, all of which have a crucial bearing on the way we respond to our current ecological crisis.

As Jeffrey Summit puts it elegantly in his research on Jewish congregations:

> I found again and again, when these nonspecialists spoke about music in ... worship they were in fact talking about the deepest spiritual questions in their lives. What tunes and chant represented the essence of who they were and what they believed ... ? What music constituted authentic practice? What was their relationship to their ethnic and religious history? Where and when did they feel truly comfortable and fully at home. In my many conversations and interviews, we spoke about music, but the real conversation was about the locus of core meaning in their lives. (2000, p. 18)

To paraphrase, inelegantly, Summit's sentiments in relation to music, climate change and ecology: as I interviewed a range of different people on questions of music, spirituality and the environment, I found again and again that they were in fact talking about fundamental issues regarding our relationships with the world around us, the relationship between spirituality and ecology, and the different ways in which we can respond to the climate crisis that we are facing. The research began as a somewhat personal journey of exploration and discovery but, as I explored, I became more and more convinced that the perspectives I was coming across needed in some way to be shared. Many of the groups and individuals were trying out new experiments on their own, with little sense as to what other groups were doing, and sometimes with little sense of what really might be the best way forward. I decided I should try and write about the experiences I was encountering, both to help make the diverse range of activity that was emerging visible to those finding their own musical or not-so-musical paths through a time of environmental catastrophe, and as a chance to reflect upon some of this activity from a critical standpoint – to ask what different projects and avenues do or don't achieve and to reflect both on what might be emerging and on what might need to emerge.

This book is my attempt to bring together some of the voices, stories and perspectives that emerged over the course of the different conversations that I had, bringing them into dialogue with recent scholarly research on a wide range of related areas and topics. Through these different conversations I explore the role of ritual, music and worship in the face of the current climate crisis, and I suggest that all of us – musical or not – have something to learn from the different responses of these individuals and communities to the crisis we are facing. They illustrate something of the range of issues and questions that we all, in some way, need to grapple with, and help us to think through the different kinds of response that are available to us in community. Before turning to the different voices of the people I spoke to, however, it is important first to offer a sense of context, and to take a look at some of the existing conversations that have taken place when it comes to ecology, music and religion.

Christianity and the environment

The relationship between Christianity and the environment has evolved considerably over the course of recent decades in response to critiques from other parts of society and as a result of changing patterns of awareness in society more broadly. In her exploration of the gradual greening of religion and the ways in which different Christian networks have grown up to pray and campaign around environmental issues, Maria Nita (2016) has done important work in tracing some of the different forms that both environmentalism in general, and Christian environmentalism in particular, have taken over the course of the twentieth century. Nita describes the way in which critiques of Christian relationships with the environment in the 1960s and 1970s led to the development of early Christian eco-theologies. The suggestion that Christian emphases on a transcendent theology and human–nature dominance were partly to blame for broader environmental destruction led Christians to respond in a number of different ways, one of which was to try and recover a more ecological reading of Christian traditions, readings that moved away from dynamics of dominance and sought to ground themselves more closely in a relationship with the natural world. These early attempts to develop a more ecologically sound theology led, in time, to a more sacramental perspective in the 1980s and 1990s, focusing on the divine grace present in creation as theologians sought both to respond to new scientific discoveries and to offer a more comprehensive lens on the Earth and the natural world as a whole. An eco-theological tradition developed that tried to weave together Christian faith and the environment in new ways, informed both by contemporary sociocultural developments and by the resources present in Christian traditions themselves.

Theological reflection on the environment has taken many different directions over time and, while the vast blossoming of research on climate change and ecology makes comprehensive documentation a somewhat daunting task, Celia Deane-Drummond (2017) has helped to map out an overview of the contemporary landscape, offering a useful summary of some of the key trends in recent eco-theological thought. While Maria Nita traces the development from an apologetic to a sacramental perspective, Deane-Drummond traces a similar movement as theology has shifted between an anthropocentric (human-centred) perspective, a more biocentric (life-centred) perspective, and a theocentric (God-centred) model that seeks to bridge the two. A historical focus on creatures as resources for humans to use has slowly developed into an appreciation that they are more than this and that there is the potential to integrate different priorities through stepping back and taking a more comprehensive theological standpoint. Other streams of thought have had important influences on Christian reflection, and Deane-Drummond traces the influence of ecofeminist perspectives that have shown how our relationships to gender and the natural world have often been bound up together in hierarchical and oppressive ways. These can be seen in our imagination of Earth as mother, in the idea that women are particularly close to nature, and in our consequent imagination of both in more passive and exploitable roles set aside from what are often imagined to be more masculine spheres of rationality and domination. Changing Christian perspectives on faith and the environment owe an important legacy both to ecofeminist work and to longer traditions of ecocriticism more generally, as they have helped to show the ways in which human–nature dualisms and hierarchies have fostered alienating and oppressive relationships between human beings and the rest of the natural world.

Eco-theology is not simply a matter of critique, however, and Deane-Drummond also describes more creative movements that have emerged in recent years: a North American movement, for example, that attempts to reimagine the grand cosmic narrative of creation in light of scientific stories of the universe; a trend to think about ecology in terms of flux rather than more classical models of equilibrium; and a whole set of new constructive theologies that seek to understand God, the Scriptures and key aspects of systematic theology through an ecological lens. This flourishing of different thought and scholarship can all be brought back to a central set of issues. It seeks to rethink problematic Christian attitudes to the environment and to reimagine the world in a way that does better justice to all living creatures. Through ecological reflection, theologians have begun to rethink structures of power and domination and our imagination of them in relation to God and the cosmos. They have increasingly sought to listen to marginalized theological voices and communities, and they have begun to re-examine longer-standing theological interpretations, commitments and traditions in light of our changing awareness of our ecological situation and responsibilities.

Theology, of course, must lead to action, and Maria Nita draws out the links between eco-theological thought and activism, the opportunity that eco-theological approaches offer for the renewal of faith, and the way that eco-theology often participates in media discourse and dialogue, forming part of a broader socio-religious intersection. Nita traces more activist

Christian environmentalist initiatives back to around the early 1980s (2016, p. 28). She describes the setting up of the first A Rocha conservation programme by Peter and Miranda Harris in 1983 and the John Ray Initiative, which focuses on education, research and advocacy in 1998. Alongside these, Nita notes the gradual turning of NGOs such as Christian Aid, Oxfam and CAFOD to environmental concerns in the years after their founding, the growth of networks such as Green Christian and Operation Noah alongside the Creation Spirituality and Green-Spirit movements, and more recent developments such as Forest Church and Eco Church, which are increasingly on the radar for a range of different congregations and environmentally interested Christians. While many of these initiatives began at the margins, they represent a gradual turning of segments of Christianity to ecological activism and the gradual realization that ecological crises often necessitate associated changes in religious engagement, even when the largest impact begins with a dedicated few. In describing these longer trajectories, Nita observes the different ways in which Christian ecological concern has translated into the activities of networks engaging in ritual, prayer and protest in recent decades. While these have often been primarily the domain of activists rather than something broadly adopted, crucially Nita suggests that these movements now have the potential to move from a peripheral role to become the future of the different traditions out of which they have emerged. We are clearly at an interesting moment of flux and change. Before we look at recent change and innovation, however, I want first to turn to the second theme at the centre of the different projects I stumbled across, namely to questions of music.

Music and ecology

As a number of different authors have discussed, a developing awareness of ecological disaster has made itself felt in recent musical trends as much as in the realm of theology. Musicians in a broad range of settings have been active for environmental causes for some considerable time but, more than that, music has often shaped imaginations of the natural world in different ways and in different contexts regardless of any explicitly activist orientations. While musicians have responded in creative ways to the changing world that they see around them, within scholarship of music the field of ecomusicology has grown in recent years to acknowledge the broad range of intersections that different musical practices have with environmental concern on a range of different levels. Exploration in this area can focus on topics as diverse as imaginations of natural environments in American popular music (Ingram, 2010); the compositional experiments of ecological sound artists (Gilmurray, 2017); the use of music as an integral component of environmental activism (Pedelty, 2016); the musical dimensions of birdsong (H. Taylor, 2017); the sound-worlds of indigenous peoples and their integration of human and non-human sound sources (Feld, 2012); or the environmental impact of music consumption (Devine, 2019) among many, many others.[1]

Within this developing ecomusicological awareness, Christian musical traditions are as ripe for ecological exploration and development as any others. The history of Christian musical engagement in this area is far from monolithic in character, and a wide range of different attitudes, priorities and starting points have emerged over the course of time as different individuals and communities have responded to the diversity of situations and contexts that have surrounded them. A quick online search for Christian environmental songs or, indeed, a glance at the thematic index of a songbook or hymnal, swiftly reveals what can easily be taken as a kind of background status quo. Individuals, organizations and committees have, over the years, sought to curate selections of repertoire devoted to the natural world, to creation and to environmental themes. These range from somewhat predictable lists of old favourites to more adventurous offerings that seek to include songs by composers who users may be less familiar with. Often Christian environmental charities seek to offer some resources or information.

Operation Noah (2014), for example, make available a relatively short list of 12 songs including well-known offerings such as 'All creatures of our God and king', 'For the beauty of the Earth' and 'How great thou art' alongside late-twentieth-century classics such as 'The servant king', 'Beauty for brokenness', 'Think of a world without any flowers' and a few lesser-known and more reflective options. While Operation Noah's list is somewhat on the short side, the compilers refer their readers to a more comprehensive listing compiled a number of years ago by Eco-Congregation Scotland (2012), a movement of congregations seeking to address environmental issues in the context of congregational life and missions.[2] The Eco-Congregation list is somewhat longer, incorporating nine new hymns made available online, 12 older but lesser-known hymns by the English writer Jan Struther, a list of well-known anthems suitable for liturgies focusing on creation or environment, and 100 or so listings of songs that might be found in churches' existing hymnals and which in some way reference the Earth, the world, creation or elements of nature. Similar resources can be found on Methodist websites focused around *Singing the Faith*, when browsing by topic on the online hymn database Hymnary.org, at the Calvin Institute for Christian Worship, Green Christian, the Climate Sunday website and a number of others.

Looking at the kind of repertoire that makes it into these lists, it quickly becomes apparent that some songs are written specifically in response to ecological concerns, while others simply integrate imagery of nature or the world for a range of other purposes. This means that, while some of these songs easily feel suitable for a time of ecological crisis, with others the connection is less direct, and we are quickly faced with the question of whether or not their message is the one we need to hear in the current moment. Chris Tomlin's well-known worship song 'How great is our God', for example, makes reference to the Earth rejoicing, bringing some planetary imagery into the lyrics of the song, but it does so largely in order to emphasize the splendour of God and the scale at which he deserves to be praised. Well-known hymns such as 'All things bright and beautiful' operate in similar ways, enjoying the beautiful harmony of the natural world around us and drawing upon our delight in the beauty of creation to remind us of the place from where it comes. This is certainly ecological writing; however, in the face of recent ecological crises it can sometimes feel a little pre-critical or naive, painting a portrait for a more optimistic time before we understood the full scale of ecological catastrophe that we now need to face up to. Over and against popular theories that it is often the best of past hymnody that survives into present-day collections, it should be relatively uncontroversial to suggest that the repertoire that survives or that has achieved a certain level of status and success is usually a result of particular theological/musical interests, certain kinds of usefulness, general popularity, the deployment of power and a whole range of other factors that are rarely quite so single-dimensional in nature. The current ecological crisis has not historically been a major driving factor in the selection and composition of repertoire for hymnals, and that means that we cannot necessarily expect the resources available to us to speak directly to current needs and requirements.

Understanding some of the different historical moments that have shaped the current repertoire helps us to understand some of the forms that it has taken. In a relatively short summary piece in *The Canterbury Dictionary of Hymnology*, the hymnologist Gillian Warson traces some key authors and trends in relating Christian hymnody to the environment. She describes, for example, the way in which Charles Wesley contrasts the kingdom of God with the vagaries and disasters of the natural world, John Austin's usage of natural imagery alongside an 'air of happy innocence', the ways in which writers focus on the 'innate goodness of a God who had brought the whole universe into being' (n.d.), and Victorian innovations such as the Harvest festival and the anti-vivisection movement, with hymns that emphasized the need to live in harmony with creation. She traces sea imagery in eighteenth- and nineteenth-century hymns, which use both its danger and sublimity in literal and metaphoric ways, before moving on to hymns that respond to industrialization and urban environments. Warson also brings in attempts to focus on the environment in a more holistic manner, such as Fred Pratt Green's 1973 contribution to *Sixteen Hymns on the Stewardship of the Environment*, a work that begins explicitly to draw in issues of pollution and human dominance over the natural world. Warson concludes her narrative by describing a sense of pessimism in late-twentieth-century hymnody alongside occasional attempts at hope before expressing a wish that hymn-writers will continue to pay attention to the connection between the gospel and environmental concerns.[3]

Extending beyond the limits of Warson's narrative, we can see this tradition of engagement with ecological themes through hymnody continuing into the present day. June Boyce-Tillman (2022), for example, has sought to engage with ecological theology in her hymn-writing alongside wider themes of mysticism, feminist spirituality and inter-faith engagement. She brings a progressive voice into dialogue with a longer tradition and looks for reorientation and renewal through dialogue with different traditions, figures and theologians. Similarly, David Coleman from Eco-Congregation Scotland continues to produce new hymns as a means of 'mobilizing people's enthusiasm' for ecological issues (Interview with the author, 12 February 2021). Coleman's work attempts to fill the gaps that earlier repertoire fails to address, combining the security of a familiar tradition with less familiar ecological imagery in order to strategically meet congregations at a place where they are comfortable and move them into new places of action and engagement (2021).

While explicitly environmental songs rarely feature in the most-sung Christian repertoire worldwide, it is also possible to trace examples of songs expressing Christian care for the environment on the edges of the CCLI (Christian Copyright Licensing International) database. For example, a 1991 children's song by Paul Field and Ralph Chambers listed in the database begins with the statement 'Don't know much about the ozone layer, Rain forests seem miles away' (CCLI, n.d.[b]), before emphasizing that this is God's world and we can all do something to save it, while a 2007 hymn by Andrew Pratt begins, 'The footprints where your people tread have marked and marred the Earth, the global warming that we dread has shadowed us from birth' (CCLI, n.d.[a]). Where these themes haven't formed the centre of Christian faith and worship, there have almost always been individuals working on the fringes trying to bring them to awareness, whether it be through singing in school assemblies, through musicals, through hymn lyrics or through another medium where they've found the room to spur the interest of a particular group.

A moment of change

None of this, of course, is static, and both faith-based and music-based climate engagement have undergone dramatic shifts in recent years as wider social movements have sought to respond to the growing urgency of the global climate crisis. Indeed, we could say that they have slowly moved from the fringes to become more mainstream phenomena precisely because the broader climate movement has gained increasing visibility and momentum over the course of the last decade. Movements such as Fridays for Future and Extinction Rebellion participate in and serve to foster new social dynamics that build upon earlier developments but which do so with new emphases and with their own distinctive characteristics. They have begun to respond to the growing urgency of the current moment, and they have done so in a way that hooks into contemporary modes of social protest that are highly networked, globally connected, democratically conscious and (social) media aware from the ground up.[4]

A new wave of social activism does not, of course, in and of itself mean a great deal for Christian groups. As with many other social actors, churches and Christians are quite capable of allowing broader trends and developments to pass them by unaffected and unmoved. In this case, however, it does seem that a broader wave of activity outside of the church has been met with a series of movements within Christian communities as they too have become caught up in the broader attention that climate issues have received and the broader movement to engage with them as an urgent global and societal crisis. Importantly, these issues are, in many places, no longer understood as highly partisan in nature or the preserve of a particular ecologically conscious interest group, but as planet-wide in nature, affecting all people and places and demanding some kind of action across the board. As a result, a much wider range of groups are now contemplating their own mode of engagement, sometimes finding resources in their own history in order to do so, and sometimes finding that those resources seem peculiarly inadequate to the changes with which they are being confronted.

If we turn to some of my early fieldwork interviews, we can see the way in which a need for reinvention often arises precisely out of this kind of re-evaluation of practices and resources in the light of changing social contexts. Mark, a vicar and ecological activist involved in Christian Climate Action, made clear the struggle that can take place to reconcile existing resources to the changing world around them:

> We're all struggling, and I'm also aware that for people in congregations you're getting so many mixed messages, aren't you? You've got traditional hymns with a sort of very safe universe, you've got the psalms saying sometimes, you know, God is in his heaven, everything is well, and you've got scriptural messages which, of course, we know all say different things at different times. You've got Victorian hymns with their particular empire perspectives, you've got the trendy vicar talking about the risk of extinction … We are confused, and it must be confusing, but that's putting it negatively – putting it positively is there's like a storehouse of resources and images and scriptures and phrases and stories, historic and prophetic and contemporary. (Mark, interview with the author, 9 January 2020)

A sense of confusion and dissonance can emerge where re-sources that were made for one context find themselves in a new situation. Different messages and narratives come together and the sense of how they might cohere isn't initially obvious. Perhaps there is some possibility of delving into them and sort-ing something out, but it seems a new sense of coherence is unlikely to arise on its own. Nancy, who has also served as a pastor, highlighted how the specificity of a particular situation can sometimes demand repertoire that is tightly focused on current events and which addresses itself to dynamics that are at the forefront of current concerns and trajectories:

> If you look through hymnals, you know, you're not going to find songs on this, I mean you'll find general creator songs, yes, but I mean not real specific to – like you said – to the ecology, to things that are happening in our world, the lack of support for climate change and you know how is creation waiting, how is creation groaning and to actually hear the groans of creation. (Nancy, interview with the author, 14 January 2020)

This concern can extend further to the feeling not just that the existing repertoire is too general but that it isn't suitable at all. Alison, who has spent a long period of time developing nature-centred ritual and musical practices, explained to me that there was simply nothing at all that was suitable for the kind of nature-based rituals that she was interested in developing:

> And so obviously when I was looking at creating ritual ... I'm a liturgist too so I wrote a lot of the rituals that we did right at the beginning ... I'm looking for songs and there aren't any, so basically it was just a needs must, you know – I needed to, and I'd been writing worship songs for years and years, so it's just a case of moving on and writing ones. (Alison, interview with the author, 21 January 2020)

In the light of sometimes confusing or missing repertoire it makes sense to innovate, to change, to adapt and to go about shaping songs, rituals and practices which feel right for the current ecological moment, that put the emphasis where it is currently needed, and which respond to the realities by which we are currently surrounded.

The meaning of innovation

In the face of pressure in some way to innovate, it is worth exploring, at least briefly, what the idea of innovation might entail. The philosopher Boris Rähme has drawn attention to the complex and contested relationship between religion and innovation, highlighting the diversity of approaches and value judgements that surround it (2021, pp. 310–13). What one person sees as innovation, to another will have a longer lineage or parallel. Likewise, while one person may see innovation as the brave efforts of individual pioneers, another will see it as a slow process of institutional evolution or the result of changing circumstances. How, after all, does Christianity innovate? By radical breaks and reformations? By splits and schism? By the gentle guiding of the Spirit? Or in a long continuity with tradition that nevertheless requires a degree of calibration to the world around it? Does innovation come from lay people or from their leaders? From the inside out or the outside in? Should Christianity stabilize into recognizable institutional forms or are these the antithesis of the church's calling? These are not just practical, but polarizing theological questions, and they have no simple answers. All of these forms of change have in some way been seen in the history of Christian communities, and we cannot simply say that one kind happens and another not; rather, we see them all present in different places at different times and for different reasons.

In order to untangle some of this fuzziness and confusion, Rähme offers a helpful outline of different scholarly perspectives on how innovation can be understood. He draws attention to the work of Steeve Bélanger and Frédérique Bonenfant, who suggest that:

> [Religious innovation is a] collective process which, out of a will and/or desire for change in the face of a situation considered as not, or no longer, meeting current needs or aspirations, introduces religious novelty and leads, by negotiation or imposition through a network of communication, to significant, effective and lasting socio-religious change in practices and/or systems of meanings. (Rähme, 2021, p. 315)

So far, so good. But a little more is needed to understand the different shapes this innovation can take. In order to do so, Rähme points to the work of Trine Stauning Willert and Lina Molokotos-Liederman, who distinguish between five types of religion innovation: purist innovation, which aims to return to an original authentic state; strategic innovation, which embraces political or economic reforms in order to strengthen a community's political or cultural standing; adapting innovation, which responds to changes in the social, political or physical environment; unintentional innovation, which arises from the influence of external source; and emancipatory innovation, which creates spaces of agency for members (Rähme, 2021, pp. 316–17).

With this, we are given a broad perspective on what innovation can look like. Innovation need not always be exclusively forward-looking in nature but, rather, can involve the creation of change precisely in order to latch back on to something from the past. It is often contextual in nature, relating in different ways to changing circumstances around a community, but it can also arise from a desire to reform internal dynamics. Some change is highly intentional with the intention of creating something new, but sometimes innovation happens simply because a community is carried along in a particular direction. When it comes to matters of climate change and our changing ecological relationships, we are thinking of highly complex local and global systems, social dynamics and ecosystems and it is, therefore, unlikely that change will be easily restricted to one particular variety. Rather, we are likely to see a range of different kinds of change as different individuals and communities interact in varying ways with the changing world around them and draw on a range of different internal resources, histories and processes in order to do so.

Roles and movements

Innovation is not simply a matter of new repertoires or practices; rather, in the context of the climate crisis, innovation is about broader patterns of social change. Different individuals, activities and groups have different roles when it comes to processes of change and, in the chapters that follow, we will see some of the broad range of roles that different groups can take on, and the diversity of different purposes that music can serve within them. Some researchers have already begun to trace some of this diversity – Alice Hague and Elizabeth Bomberg, for example, have drawn attention to the different roles that faith-based actors (FBAs) are taking on as intermediaries between state actors, policymakers, citizens, individuals and community groups. They suggest that groups can engage in representation, as they articulate the views of their members into the wider policy discourse; mobilization, through building community and galvanizing this community into action; and aggregation, 'translating and scaling knowledge to make it relevant in different contexts' (2023, p. 600). Music can hook into a similar range of dynamics, and Rob Rosenthal and Richard Flacks outline the roles that music can play in processes of education, conversion and recruitment, both aiding in processes of mobilization and serving those already committed to a cause. More than this, however, they describe a vast catalogue of different possibilities, discussing music's role in creating solidarity; in externalizing beliefs and a feeling that a movement is real; in sharing and managing emotions that affect commitment; in persuading individuals to identify with a movement; in shaping amorphous ideas into a coherent perspective; in increasing the pleasure of participation; in fostering a movement's broader acceptance; and in helping move individuals 'beyond agreement and identification with those claims (recruitment), to taking that crucial step into concrete public movement activity' (2011, p. 173). Both music and faith have distinctive roles to play in creating social change but, in these suggestions, we see a degree of overlap, in drawing communities together, in mobilizing them into action and in communicating in a way that hooks into a sense of feeling or identity.

Neither in the case of music, nor in the case of faith-based engagement more generally, is any of this potential guaranteed. Bomberg and Hague emphasize that 'religion can have a number of effects on political mobilisation, facilitating but also hindering action' (2018, p. 591), while Rosenthal and Flacks make it clear that many musicians are quite ambivalent about the role their music can serve, unsure as to whether it has any great effect. In the case of music, the dangers include the containment of political activity; the replacement of collective action by individual identification with an oppositional performer; the exclusion of those who fail to identify with the music; the presentation of a false sense of unanimity that leads to the avoidance of important discussion around areas of disagreement; the potential for music's emotional power to persuade on a purely emotional level rather than as the result of deeper cognitive conviction; and its potential to reinforce existing power relations by encouraging accommodation to the status quo. The key question, according to Rosenthal and Flacks, is whether or not musical activity succeeds in 'present[ing] or preserv[ing] some sense of an alternative way of life' (2011, p. 194), and it is notable how easily this question can be transferred over to the realm of religion, where the potential to create an alternative vision of the world is precisely the challenge at the heart of a great deal of religious activity. The authors do not prescribe a particular recipe for success, suggesting that many of the factors that make a difference are contextual and beyond a musician's control. They do, however, point to the importance of dynamics of participation, of connection (or conscious break with) tradition, of musical qualities, and of examining the role of musicians. Musicians can face dilemmas in navigating the different possibilities that they face, but Rosenthal and Flacks suggest that 'rather than argue about whether there are intrinsically superior forms of political art, we need to recognize that different musics and different approaches are more and less useful in different times and different stages of a movement, as well as to different individuals with their variety of relations to a movement' (2011, p. 248).

Turning to fieldwork

Ethnography and qualitative fieldwork are key tools in understanding the way in which change and social movements function, and ethnographic study of faith-based ecological initiatives has slowly gained ground over recent years. Sarah McFarland Taylor (2007) has used ethnographic fieldwork to examine the ways in which Catholic sisters blend faith and environmentalism. Fieldwork enables her to paint a detailed and multidimensional portrait of different communities and their integration of ritual, contemplative practice and practical projects as they reinhabit their faith and the land around them. Katharine Wilkinson (2012) focuses on evangelical movements in the USA, examining the different twists and turns that take place as different leaders and groups navigate conflicting loyalties and influences. She describes the ways in which change is strategized and the ways green and evangelical logics both stand in tension and come together. Other researchers have begun to see the need for similar work on this side of the Atlantic. Alice Hague (2018) has undertaken environmentally focused ethnography in three Edinburgh congregations, examining the way in which theological motivations and practical factors come together in the context of church community life. In a journal article co-written by Hague together with Jeremy Kidwell (2018) and a number of other researchers, the authors emphasize the way that ethnographic work is able to go deeper than survey studies and other less-engaged research methods. They examine Eco-Congregation Scotland and the way in which process, structure and action come together in what they describe as a kind of eco-theo-citizenship which integrates belief, action and citizenship in a mutually reinforcing spiral.

The rest of this book builds upon the conviction of these different researchers that fieldwork has important potential in understanding the richness of different movements and communities and the realities that they inhabit, bringing together dimensions and insights that are not so easily available through other means of investigation. In exploring musical innovation through ethnographic methods, the different chapters draw out conversations with a range of different communities and individuals engaging in practices and creative processes which, in some way, do something new. This interest in change and innovation is a result of my own observations that innovation is currently taking place and of a deeper-lying conviction that the changing world around us demands some kind of change. This means that the chapters ahead deal with uncertainty, and with the trying out and testing of different possibilities. They are based on a sense that practices of faith are far from static but always in a state of flux. Music is a highly visible medium, and musical creativity offers a helpful focus point around which a range of different thoughts, feelings, relationships and interactions are able to crystallize and come to expression.

Many of my encounters with different projects came initially through social media. The proliferation of live-streaming and video sharing has made social media into an increasingly popular medium for sharing musical projects and events. Music takes place in a variety of different geographical locations, but it rarely remains restricted to them, spilling over into online forums for wider awareness and circulation. More than that, however, since many of the projects that I investigated have some connection not just to local communities but to broader faith-based networks, social media is often a place where individuals connect to broader communities of practice. They use it not just to publicize events, recordings and products, but to seek out advice, to organize together, to gain inspiration and to reach out beyond the smaller pools of like-minded individuals that are available in their immediate geographical neighbourhood. Forest Church practitioners share many of their experiences with one another through an online Facebook group; Christian Climate Action maintain their network through a highly active WhatsApp group; and Resound Worship co-ordinate their network through a website and different online forums and media. Local groups are far from unimportant; indeed, they are crucial for much of the creativity that I discuss over the course of the coming pages. However, this local activity is often made visible to and inspired by these broader networks. It is therefore fundamentally hybrid in character, situated at the intersection of networks and communities that are physically present and further away. It is also often somewhat fragmentary in nature, existing in little pockets of interest that are embedded in communities and situations where ecological music is not the primary focus of the group as a whole. This means spreading a wide net in order to trace different projects and activities rather than necessarily following one particular community in depth over a long period of time. Since my initial work on this project coincided closely with the start of the Covid-19 pandemic, much of my own contact with different interview partners was also through different online media, writing emails, and WhatsApp and Facebook messages before setting up Zoom links for conversations. Through these different conversations I entered into the worlds of different individuals and communities, learning about their thought processes, experiences and feelings. In contrast to my own previous research projects, I focus here on the process of creative activity more than on the experience of its results; however, these boundaries are fluid, and I hope it quickly becomes clear how the connections I explore open out to myriad wider worlds beyond them. Indeed, the global crisis we now encounter demands that they do so.

In what follows we are going to encounter a range of different groups and activities, each of which has a different role to play in the face of our changing environmental relationships. Each draws together a different set of concerns and priorities and shows a different facet of the engagement between Christianity, devotional practices and the environment. In bringing them together in the pages of this book I hope to paint a picture of the diverse possibilities that different individuals are currently exploring, while setting this in a broader context that helps to understand the different value they each have in facing the ecological challenges and changes by which we are currently surrounded. We enter the projects through a focus on music, but this serves as a gateway into much wider debates and issues. With this in mind, we turn now to the first project that I encountered right at the beginning of this research, back in 2019.

Notes

1 See Allen and Dawe 2016 for a selective sample of some of the range of work that took place up until the mid-2010s. A more recent article by Mark Pedelty et al. (2022) helpfully summarizes some of the many different strands that have emerged over the course of recent decades.

2 In reading through the draft of this Introduction, the current chaplain of the Eco-Congregation was keen to emphasize that this movement has evolved considerably since 2012, both in terms of its functioning as an institution and in his own understanding of the theological reach and implications of the ecological crisis.

3 Other authors offer a range of perspectives on the different shapes that landscapes of Christian spirituality of the environment have taken over the course of the centuries. Alan Hall (2014), for example, traces a concern for God as creator back to early hymnody, following various developments through deist celebrations of a mechanical universe, romantic nature poetry, a focus on God as creator and redeemer and twentieth-century compositions that focus on problematics of power, human destructiveness and environmental concern. A number of different writers emphasize that some kind of interest in the natural world or even ecological concern is by no means a new phenomenon and there have been numerous Christian impulses over the decades and centuries to connect together liturgy, devotion and the world around

us. Brett Malcom Grainger (2014), for example, has sought to recover the 'vital landscape' of evangelical religious practice in eighteenth- and nineteenth-century America, arguing that, as early evangelicals often met outside, they often imbued the natural world around them with layers of sacred meaning, laying biblical names and location over familiar landscapes, and associating particular places with spiritual phenomena or experiences. In a similar manner, Leigh Eric Schmidt (1991) has traced different nature rituals and liturgies associated with American Protestantism from nineteenth-century Arbor Day celebrations through early twentieth-century Bird Days, Flower Days and other celebrations and outings to Earth Day, the Environmental Sabbath and a range of other late-twentieth-century practices that foregrounded the ecological crisis and sought to respond to it through special events and liturgical acts. David Kendall goes further back than any of these authors, suggesting that we return to consider the writings of the philosopher Boethius and sixth-century Christian concepts of *musica mundana* and the music of the spheres in thinking about Christian musical relationships with the natural world around us. Kendall proposes that:

> ... early thinkers were working within an ecotheological concept of music, an integrated and interrelated universe in which the proper use of music was indicative of, and a way to maintain, a properly balanced cosmos in all its celestial, spiritual, human, nonhuman, and environmental aspects. As such, music became an integral part of creation and how created beings relate to their various physical and conceptual environments. (2016, p. 120)

4 A number of different researchers have sought to understand the current climate movement and the particular connections and histories that serve to set it apart from early modes of environmental activism. Christopher Chase-Dunn and Paul Almeida, for example, connect the contemporary climate justice movement not just to longer-standing environmental activism but to the global economic justice and anti-war movements of the early 2000s, and to the development of a global internet, global neoliberalism and free trade. The model developed through these early 2000s movements offered a template for coordination on a global scale (2020, p. 161) that differs in important ways from earlier activist movements. Joost de Moor et al., meanwhile, trace four kinds of novelty within recent climate actions, focusing on themes of large-scale mobilization, disobedient action, speaking to local and national governments and listening to the science (2021, pp. 623–4). Peter Gardner, Tiago Carvalho and Maria Valenstain (2022) trace a greater, but still limited, transnational dimension that has arisen since 2018 but also draw attention to the importance of major protest events in spurring

new involvement. Anna Friberg, similarly, focuses on particular events, but traces them back a little further to the 2009 United Nations Climate Change Conference in Copenhagen and, in particular, the increased media attention and increasingly visible NGO engagement that emerged around this time (2022, p. 3).

I

Worship and Climate Albums

It is September 2020, and while some lockdown restrictions have been eased, many limitations are still in place. Throughout the early months of the Covid-19 pandemic I have been tuning in to the Diocese of Oxford's Sunday services on Vimeo, enjoying the diverse range of Anglican traditions they attempt to bring together and appreciating the more ecumenical experience that the digital format seems to enable. To celebrate the season of Creationtide, the diocese has put together a service focused explicitly on themes of climate and sustainability. We begin with a short documentary film about an Extinction Rebellion protester before the opening of the liturgy, and this leads directly into a welcome from the Bishop of Oxford, standing in front of a communion table ready to celebrate a service of Communion. After some brief introductory words, the bishop introduces the first song of the service:

> As we hold in mind our stewardship of the whole Earth, the beauty of creation and the extraordinary challenge we face in our generation to see climate change reversed and this good Earth preserved for future generations, we hold in mind the poor across the Earth disproportionately affected by climate change, and all who are working together for the common good in this part of our mission.

A relatively sparse beat fades in, accompanied by gentle electric guitar figurations, as grey images of dry and cracked landscapes appear on the screen. The first song of our service is not going to be a song of celebration or praise but, rather, a newly released track from the album *Doxecology* meditating on whether parched and treeless fields are still able to cry out to God and pleading for God's mercy to forgive us and renew the world around us. It is a stark opening to the service, but one that sets us up well for the confession that immediately follows. Having recently listened to the track for the first time on YouTube, I am impressed by how quickly after release the song has already been incorporated into this kind of celebration, noticing how clearly and obviously it connects to the wider project which the Church is attempting in their climate-focused liturgy.

Popular music and the environment

In trying to figure out why and how different groups were engaging with issues of the climate through music, the first project I sought to establish contact with towards the end of 2019 was a climate-album project from the songwriting collective Resound Worship. As one of the most-publicized climate music projects in recent years, this seemed an ideal place to start, and one that would help me get a grip on some of the different ways that people were thinking in writing music for the climate crisis. Resound Worship is a website and collective of songwriters with a broadly evangelical base. It focuses primarily on music in the contemporary worship music genre, and thus combines a focus on popular music styles with an interest in the needs of devotional environments. The website encourages individuals in different ways to write songs and provides various kinds of resource to enable this. The group's best-known activity is the 12-song challenge, which encourages musicians to write a song every month over the course of the year in response to particular prompts and suggestions. In writing regularly and communally, the idea is to develop skills and locally sourced repertoire while enabling participants to encourage one another in the process of creativity. In 2019 they launched a project entitled *Doxecology*, a search for new songs written around themes of creation, the environment and ecology to which individuals could submit different compositions, with the hope that a selection of these would eventually be compiled together in album form.

Environmentally concerned popular music is far from something new. In *The Jukebox in the Garden*, David Ingram (2010) describes the different ways in which artists from almost every genre, from folk, blues and rock through to R&B and hip-hop, have engaged in some way with environmental themes. Each genre affords different possibilities for expression and action as a result of the different attitudes and social positioning embedded in different groups and musical forms. Blues and country, for example, conjure up a particular imagination of rural spaces, and deal with issues of race and class – they often describe a sense of struggle against the challenges thrown up by nature alongside a sense of nature as a nourishing home. Folk music has more activist tendencies as a result of its connection with the political left, while early rock can tend to embed a more mystical perspective on the natural world as part of its wider countercultural relationships. Amid a broad range of different possibilities and options, Ingram traces common themes across the different musical genres that he surveys, suggesting that popular music tends to approach the environment through lenses of elegy or of satire, highlighting contradictions between actual reality and the ideal. It can conjure up visions of a natural world in which humans feel at ease and at home, it can meditate on loss or it can stand in the space between, denunciating those responsible for that loss (Ingram, 2010, p. 52).

We should not necessarily expect broader patterns from popular music more generally to automatically carry over to the realm of worship, but we should also be wary of assuming it to be completely exempt from general tendencies. The tensions between Christian popular music and a more secular mainstream are well documented, and Christian artists have adopted a range of strategies to negotiate the different pulls of faith and performance (Howard and Streck, 1999). The worship music genre, which quickly came to dominate Christian popular music output, is perhaps the genre that is most distinct from secular output, focused as it is on an act of divine worship.

So how do you go about writing a climate-focused worship album? And what are the individuals involved in writing one trying to achieve exactly? Should we expect them to navigate environmental issues in a similar way to other writers of popular music? Or do they carve out their own set of distinctives specific to the worship music genre? In approaching different songwriters and musicians for interview, the answers to these questions were far from obvious to me. I had little idea what a Christian climate album might sound like or what it might include, and I was keen to find out more.

Challenges of genre and tradition

The *Doxecology* project itself has a very pragmatic focus on creating songs that might be useful within existing worship communities. One of the key questions to be faced in starting the project seems to have been how to introduce a focus on ecological issues within the limitations of the worship-song format where it might not always be the most obvious fit. How, in other words, to integrate new concerns with the existing patterns and expectations of individuals and communities. As a result both of its relation to a popular music mainstream and of its longer development in evangelical and charismatic worship environments, the worship-song format has particular generic characteristics and expectations. The form has crystallized, for example, around the use of a verse–chorus structure as a standard popular music format that enables an easy sense of orientation and repetition. As part of its lineage in experiential-centred charismatic traditions, it tends also to focus on the relationship of the individual or group to God, and the need for the singer to address themselves to God in a direct and feel-ingful manner. Joel Payne, one of the founders of the project, suggested that they tried out a number of different approaches before finding something that seemed to work:

> We wrote nine-verse epics because we had so much to get off
> our chests ... this kind of thing where you name every single
> kind of 'We don't do enough recycling, Lord, we spurn the
> gifts you've given, we burn', we had to sort of get through
> that process ... and say no one's going to sing these. (Joel,
> interview with the author, 2 December 2019)

The challenge with these 'nine-verse epics' is particular to
this format; within a hymn structure, writing that enumerates
climate issues in significant detail might easily find a more natural
home; however, the worship-song format is not generally used
for articulating large quantities of information. It is not simply
on a textual level that engagement with environmental themes
contained potential generic tensions, however; the desire to
engage with an existing mainstream also manifests in intentional avoidance of other potential avenues:

> What we don't want to be is, oh yeah, these are the eco songs,
> and because they're eco songs they're going to be folky ...
> because that's more close to Earth and nature ... we want
> people to say 'Oh, this is a great collection of mainstream
> worship songs that happen to be on ecological themes' as
> opposed to 'This is a great collection of ecological worship
> songs'. (Joel, interview with the author, 2 December 2019)

The oft-established, even stereotypical, connection that is made
in popular imaginations between folk music and nature or the
natural – as well as the different critiques that might be made
in this regard – has been addressed by commentators such as
Simon Frith (1996, p. 40) and David Ingram (2010, p. 48),
and Joel's scepticism here is well in line with the views of
such commentators that these connections are very much constructed ones and should by no means be taken at face value.
However, Joel's concern here arises largely in relation to the
conventions of contemporary worship music as a genre. The
different tensions raised in relation to generic expectations
need to be negotiated in some way in producing an ecological

worship album, and because of his desire to write songs that mainstream evangelicals will sing, Joel attempts to upset as few of the generic conventions surrounding contemporary worship music as possible.

Other writers who responded to the initial call for songs had related concerns. Rick, for example, emphasized the importance of keeping scriptural narratives front and centre:

> How do you engage environmental issues in music and worship and everything else without letting it become the focus? … in evangelical circles … if you think about the creation, fall, redemption, restoration, it was all about fall, redemption and never about creation or restoration … it's like trying to get back to saying we have to start with creation and get a good understanding of the context there and how it fits into the whole story. (Rick, interview with the author, 4 January 2020)

However, Samuel raised more pragmatic issues:

> In a congregation, if you can only roll it out once or twice a year will people be able to sing it? And I think that's the issue when you come with these very single-issue songs. (Samuel, interview with the author, 6 January 2020)

While evangelical priorities serve to guide the range of possibilities available, as Rick suggests, not everything that exists within existing Christian traditions is helpful in navigating these issues and, in approaching the project, Joel emphasizes the need to renegotiate relationships of power established in existing traditions of Christian thought:

> The community of creation idea is something that came out strongly for the writers … how have you viewed yourself in relation to creation? Do you see yourself as above it or separate from it? … where you go to any church and they'll talk about being good stewards of creation, go to the theologians

and a lot of them will say 'oh, don't use the word steward, that's terrible, that's a hierarchical thing'. (Joel, interview with the author, 2 December 2019)

An in-depth exploration of climate change and a focus on the environment begin to reveal some of the weaknesses of traditional Christian framing of these issues and, as such, the attempt to creatively engage with these themes provides an opportunity to renegotiate particular dynamics and to gain the ability to see the world from a new perspective. Indeed, in some cases it was precisely the ability to break away from some of the limitations of the existing tradition that provided the excitement in going about this project:

> I feel like these are some of my most complete songs, where I'm proudest of what I've done lyrically. That's partly because I felt more free to approach them as thematic ideas in and of themselves. I felt a freedom … in fact, an obligation not to write just typical lyrics … that to do justice to these slightly unfamiliar themes, the lyrics needed to be more allusive and artistic. A lot of worship songs feel like they're about the-intense-singing-of-my-heart's-deepest-feeling-right-now … which can lead to rather leaden lyric writing, because what anyone's feeling right now is not necessarily the most beautiful or interesting thing. These songs were freed from the obligation to fall into that genre. (Chris, interview with the author, 11 September 2020)

The tentative and exploratory attitudes of many of those I spoke to over the course of early interviews suggests that clearly established guiding paradigms are largely absent at the current stage of development. When talking with different individuals about the climate albums – or, indeed, most of the other projects I will be discussing over the course of this book – it quickly became clear that both the motivations for and the approaches to engaging musically with the environment varied significantly. Indeed, in the attempt to negotiate new territory at the intersection of Christianity and an issue that has, to a large extent, emerged within a secular scientific context, there are, to date, few well-trodden paths to take.

A range of starting points

Individuals who sent in songs to the project adopted a variety of different approaches and came to songwriting in this area with a range of different motivations. Trevor, for example, had developed a broader project over a number of years in order to integrate faith and science through song. Trevor works largely in a hymn-writing tradition rather than a worship-song one, providing new texts for existing hymn tunes focusing on themes of science, cosmos and creation. This seems to sit well alongside his desire for rational explanation and elaboration:

> For me this work is primarily to encourage people to think about the complementarity of science and faith in revealing the wonders of creation and the glory of God throughout the universe ... There are many first rate, indeed brilliant, scientists and theologians who write and speak on that ... However, it seems to me these authors have to first establish their credentials within their own disciplines before they can develop ideas about the crossover points ... By contrast, poems, songs and hymns have to capture any concept of complementarity very quickly. If they don't, they will fail in their objective. For this reason such work, if well done, should be able to make the idea more accessible to a far wider audience. (Trevor, interview with the author, 17 December 2019)

We can contrast such a rational and scientific focus with that of Lizzie, who approached her songwriting from a more mystical standpoint:

> I was thinking about Communion, and the connection between God's giving himself as Christ ... in Richard Rohr's book *The*

Universal Christ, he talks about 'real presence', about the bread … recognizing that it is the body of Christ, but in the same way that everything in creation is the body of Christ … I think he used the word 'panentheist', and that felt like a kind of place for me. (Lizzie, interview with the author, 2 April 2020)

In her writing, Lizzie draws on imagery of nature as a cathedral and prayers as leaf mould, meditating on the relationship between God and – immediate and distant – space and the relationship between that presence and one's own body. The song that she submitted to the *Doxecology* project draws on liturgical responses and embodies a certain kind of eclectic relationship to Christian traditions. At the same time, it becomes almost apophatic in its attempt to reach beyond what words can describe:

I was trying to write a song that didn't give God any gender or any name except 'I Am' … thinking about how important it is that God is Father and Son and Holy Spirit and thinking, well, those are important, but … there's a whole lot of other stuff going on that I just don't have the words for … God is so much more than I could ever grasp, but also very present, very loving and very manifest in all sorts of ways which I don't really understand. (Lizzie, interview with the author, 2 April 2020)

Here, writing a worship song is centred around reimaging the relationship between God and the created world, and in doing so it involves a reimagining both of God and of the limits of the Christian faith. God's presence in the world, for Lizzie, is not built upon a sense of boundaries between God, or Christians and the other; rather, in locating God's presence in everything, these boundaries begin to dissolve.

While Lizzie and Trevor illustrate two different poles in their attitudes towards faith and the environment, other writers tread a slightly more familiar path in-between. Hannah, for example, presented to me a song written by her children that emerged at the confluence of their broader thankfulness to God, their awareness of environmental issues and their devotion to faith:

> We were just talking about what they've been thinking about and what they've been reading in the Bible ... they have quite strong opinions and want to discuss these issues more in church, and we have a group of adults who ... still have not got the fundamental of the fact that creation is fundamental to our faith ... I actually said ... 'You guys should speak to the church around it', and ... it came up that they would write a song ... they wrote it that morning, went upstairs and sang it ... then they printed the words out and they taught the church ... I think all of the songs that they've written ... even if they're not quite as openly about creation ... they all have that common thread through them, but I think ... that's how they thought about God anyway, and how they experience him ... they just wanted to write a joyful song about why they love God ... the three of them ... just wrote down a list of things that we're thankful for, and that's where the first verse came from, because in that order Amelia wrote like moths, flowers, bugs, and bees, they all wrote lists like that. (Hannah, interview with the author, 21 December 2019)

The musical initiative from Hannah's children has an impact that extends beyond the moment of writing or singing and had clear practical consequences in their church community. The song opened up a space for conversation around ecological themes, led to potential future opportunities for a teaching series on these topics, and prompted donations to the local wildlife trust. Here we begin to see the wider set of relationships that the creative process can draw upon and feed into, in a way that can sometimes be more significant in the long term than a resulting creative product in and of itself.

The frequent human experience of encountering God through nature was raised a number of times by other songwriters in their approach to songs. This existing trope performs a ready-made interface between Christian worship and the natural world, and one that, perhaps, requires minimal tweaking to bring it to bear upon current climate issues. Building on existing paradigms is a useful strategy, and Samuel's writing also draws closely on existing worship-song models through its prioritization of the scriptural texts that stand at the heart of evangelical faith. Samuel emphasized that:

> Mostly my writing comes out of taking in a piece of scripture or a scriptural idea and working around there ... I was really fascinated by the passage in, I think, Hosea chapter 4 which talks about the land weeping, I'd not come across that passage before ... the first verse I suppose is a wonder at creation, the second verse introduces the idea that we haven't done what we should have done with God's call ... verse 3 for me is about what Christ is going to do and then about ... not to wait for that to happen, but to be involved in now the kingdom coming through us. (Samuel, interview with the author, 6 January 2020)

Samuel is keen to understand the environmental situation within a narrative of God's kingdom, emphasizing humanity's role as caretakers of God's world, and a gospel message that encompasses not just personal sin and salvation but every aspect of our lives, communities and the world around us. God will renew everything, and we are called to be a part of that renewal. As a pastor he is aware that his congregation are on board with this responsibility and awareness to varying extents, some taking responsibility for their own actions, and others relying more on the idea that God will come back and sort everything out. There is sometimes a degree of caution and uncertainty in introducing ecologically focused repertoire within worship, a sense that it might, in some way, jar with existing emphases, and that care is needed to navigate these tensions in an appropriate manner.

In the different accounts given by Trevor, Lizzie, Hannah and Samuel we can clearly see the diversity of concerns that individual writers brought to the songwriting process, drawing on their

own particular contexts, interests and experiences, and addressing themselves to particular audiences in ways that speak to
their hopes and expectations of what is possible in different
places and at different moments. Human experience is diverse
and, left to its own devices, reaches out in a range of different directions as individuals grapple with their own stories and
situations – in this case, without an existing guide to follow.

Producing the final album

While responses to the original call for songs reveal a diversity
of different approaches, the final album ended up a lot more
focused in nature, with most of the songs coming from a core
pool of Resound Worship songwriters rather than from this
broader community engagement. Over the course of 13 tracks,
we move through songs focusing on God's creation of the world;
the creation's praise of God; its groaning for a different future;
the brokenness of the world; the need for God's mercy; despair
and disorientation; God's desire for healing; and our work as
tenants of God's garden. The songs cover a range of ground, but
ultimately share a more unified aesthetic and approach. They
integrate an interest in the world around us with the narratives
of Scripture and combine an interest in creation and crisis with
relatively familiar worship-song trajectories that bring concerns
and praises to God in prayer and singing.

The project went through a number of different phases before
crystallizing into its final form. The idea initially was to produce
a number of separate EPs, which would have had more room
for a variety of approaches. This idea then gave way to the
organization of a community-orientated live-recording project
before the final idea of a studio-produced album emerged when
the community recording project was cancelled as a result of
the Covid-19 pandemic lockdown restrictions. Plans changed
as ideas crystallized, as situations evolved and as those involved
in the project slowly began to understand themselves what they
were really trying to do. The decision to focus the album on

the songs produced by a core group of writers seems largely to arise out of a sense of pragmatism. Contemporary worship music has an important pragmatic lineage, rooted in its use of popular music in order to reach out to individuals who find popular music easy to relate to. The question of what works can often take precedence over more abstract questions of what a particular genre means or conveys, or of what an ideal form of Christian music might look like on the basis of theological reasoning or tradition. Following this tradition, one of the most important questions for the compilers of the album was whether churches would indeed sing the songs that they were writing:

> When I'm writing a song for Resound, I'm definitely trying to structure it in such a way that a congregation can sing it, with a simplicity to the melody and a clarity to the lyrics. My ambition for the project is that it locks in a reference point for people who listen, so when one of these themes comes up in a sermon series they'll think, 'Oh, I know where I can get resources for that'. (Chris, interview with the author, 11 September 2020)

The question of how exactly the songs might be used in worship was one that many in this core team had to wrestle with:

> I would definitely pick the song I've written and pop that in … what I'd call a standard worship time, not just because I've written it, but because I think it's not too specific that you have to be following a talk about the environment, for example. Some of the ones on there are – they're beautiful worship songs, but they're very specific … whereas I think you could throw in this song in a standard time of worship and it wouldn't seem out of place. (Andrew, interview with the author, 15 September 2020)

There is clearly a balance to be struck between general usability and the specificity of the song content, and while different songs on the album strike that balance in different ways, the striking of that balance was an important concern in selecting what made it on to the final album. Indeed, addressing a range of use-cases, perspectives and situations was another important concern in the structuring of the album as a whole:

> I suppose when it comes to what actions people would take, I think this is why this song is on a collection of songs that cover a breadth of themes. So you probably wouldn't sing mine and then go, right, we've got to do something about this. There are plenty of songs on there which will make you think that, so I guess this is doing a different job. (Andrew, interview with the author, 15 September 2020)

> I was conscious of that song ['If the fields are parched'] as being part of this wider body of work … I love the fact that 'Be still my soul' comes after mine. In a sense, they contradict one another … one is saying everything's terrible, the other is saying don't worry … As I engaged the whole issue, I feel like there are contradictory truths in some ways, but actually the sum total is truer. (Chris, interview with the author, 11 September 2020)

At the same time, there is an awareness that perhaps there might sometimes be a need to go in a different direction from that which a more standard worship context might provide:

> I think one of the things that Sam Hargreaves again said was … 'How do you judge a successful worship time?' … in charismatic churches often we see the purpose of worship to be getting to the place of intimacy with God. And I don't think there's anything wrong with that … But then he says that isn't always the journey we need to take … Sometimes a journey we need to take in worship is preparing us to pray, or it could be preparing us to focus on a particular issue. And so just because it doesn't end in this kind of glorious moment of intimacy doesn't necessarily mean that it was an unsuccess-

ful time of worship. (Andrew, interview with the author, 15 September 2020)

We are faced with the question of diversity in worship, and the possibility that relatively well-established charismatic trajectories that move in a progressively more intimate direction over the course of an extended worship set might not always be exactly what's needed, particularly when faced with the challenge of themes such as climate change. While some themes that have the potential to disrupt established norms in worship can fail to make a significant long-term impact, this is one where the topic is so unavoidable that perhaps this might lead to a different reception than songs on, for example, social justice that have often come and gone within particular moments inside of worship music circles:

> I was going to say an interesting thing, because you mentioned like those social justice projects. And that's interesting because it's made me think it does feel a bit different with this concept album in that with those you'd get churches that were definitely into that social justice sort of thrust, and it would be a massive sliding scale, there'd be others that wouldn't be. And yet, with this, it's sort of something that you feel like you, no one can really avoid it as a topic. It's sort of vital to how we're going to go forward. And even outside of the church it's a huge thing. So I think just because of the nature of this specific niche, you know, project topic, it makes it much more palatable. (Matt, interview with the author, 9 September 2020)

Concept albums might not always be a good idea and attempts to produce worship music focused on particular themes have not always met with long-term and sustainable success. However, not all issues are the same, and perhaps the climate crisis is truly different enough from other themes that have come and gone to actually make a lasting and transformative impact. The scale of the issue means that congregations face this topic in a different way from other challenges, and this leads to a possible cause for optimism at least by one or two of those involved.

Transformation through creativity

One of the key impacts of the album was on the core team involved in producing it. Creative involvement can be a powerful avenue for engaging with a particular theme or area of concern, and it is notable that the project had a significant transformative impact for a number of the team involved. For some of the core songwriters, it was precisely through the album project that they themselves began to engage with ecological themes more seriously:

> I was not passionate about it to start with, I just kind of thought it was a bit weird thing to write into ... How are you going to write a song about recycling? ... What I quite liked about taking Psalm 104 was that it could be a bit more of a normal worship song ... And I kind of thought, 'Well, I can go with this' ... spending the time getting stuck into what the Word says about creation has been really beneficial, and I feel like it's opened my eyes to things ... Extinction Rebellion, or the big kind of protests like that – I would no longer look down my nose at that. (Andrew, interview with the author, 15 September 2020)

> My initial reaction was like, 'Oh, man, we're gonna write a load of unpopular songs that no one wants to sing'. In broad terms I was like, yes, these things are important, but the thought of spending my songwriting energy on something which seemed relatively peripheral at first glance was a little bit like, 'Oh, I'd rather write on the cross'. It was only really through the journey of reading the theologians, wrestling in the Scriptures, and writing the songs ... I was astonished at how deeply embedded this theme is in the central story of Scripture. You know, Romans 8 ... that chapter is quoted

almost more than any other in charismatic worship services, and the rest of creation is so present in that chapter. I suppose I hope that because we've written coming from that place ourselves, these songs will help other people on the same journey. (Chris, interview with the author, 11 September 2020)

For these writers it is the process of creative engagement that leads them to take ecological themes more seriously and to experience them more deeply, something that also has a knock-on effect in terms of their relationships with broader groups, whether that be through their own attitudes changing, or the hope that their journey or output will become something meaningful to others wrestling in a similar situation. The process of album production begins with their own journey and transformation, and in this sense draws less on the diversity of understandings and approaches that individuals bring from outside the group, and more on a shared journey in which a core community engages in a process of collegial discovery from a relatively similar starting point.

Responses to the project and taking it forward

The project seems to have had a wider impact beyond the writers directly involved; the album was generally well received by churches and even turned out to be one of Resound Worship's more successful endeavours. In treading relatively new territory with the album, there was a significant display of interest on the part of churches and, in particular, environmental groups:

In the last couple of weeks, I've had a high number of licensing requests. Joel has been speaking to lots of diocesan or denominational environmental officers. Writing on such an undeveloped theme means that the kind of folks who've been crying out for good resources on this subject, or hadn't even imagined that there could be good resources on this subject, are really shouting about the project for us. It's very tangible that writing into such an underdeveloped area has struck a chord with Christians who care about the environment. (Chris, interview with the author, 11 September 2020)

Indeed, Joel indicated that one of the songs, in particular, ranked especially highly:

'God the maker of the heavens' was our second most popular song in the last six months. Now, that doesn't surprise me to some extent, because … it's very accessible and you can sing it for Harvest or for Climate Sunday and so on … I would say … [congregations] probably are picking [these songs] for special services. But then I think … there's a few on there which really you can sing anytime … and they're the kind of celebrate creation-type ones. The ones that are more lamenty or confessional, they would sort of be odd to throw in, I think. (Joel, interview with the author, 2 December 2021)

The production of songs in this area ties in with a broader interest in environmentally themed services and fills a key market gap, as churches search out music for these events. Not all songs are considered appropriate for services without a special environmental focus as a regular component of Christian worship; however, in producing more generic texts alongside more specialist resources, there is the potential for the album to find multiple points of entry into Christian worship and liturgy. In this sense, it is a highly strategic product, addressed to a range of use-cases where it will be able to find a role.

The album also opened up the chance to collaborate and engage in dialogue with a number of different organizations, including A Rocha, the Church of Scotland and Christian Climate Action. It opened up the chance for mutual understanding and for putting on collaborative events that might not have been possible without the existence of the project. And the level of interest in the album carried over to a high level of interest in the subsequent tour, particularly since the timing of the tour lined up with the hosting of COP UN Climate Change Conference in Scotland, an event that inspired a broad range of Christian interest and activism across the UK.[1] The album is not seen as the end of the journey, however, and some of those involved described both their hopes for the future of the album and what they consider to be a reasonable trajectory over the next ten years or so:

> I guess my biggest ambition for the project is that it opens up space for other people to keep working on the theme. The closest parallel that I can see in recent years is with lament; 15 years ago there was a lot of talk in worship songwriting circles about 'Where are the laments?' ... and in the last 15 years, the upsurge of writing on lament has just been unstoppable. I would love for *Doxecology* to be seen in five years' time as an album that caught the bubbling up of a Zeitgeist ... for Chris Tomlin to think 'I need to write one of these songs' ... that churches will be crying out for songs like this. Right now, if I were Chris Tomlin's record label, I might be thinking, if you write on this subject 40 per cent of the people who buy our records are going to be up in arms. But I'm pretty confident that if we had this conversation in ten years' time, there will be a whole bunch of stuff ... major artists will have put out songs covering some of these themes ... because it's just not going away. (Chris, interview with the author, 11 September 2020)

There's a sense both that some residual resistance remains to including these themes in worship and that, as the world around changes, this is also going to change. The album stakes out some territory, responds to pre-existing currents and hopes to enable something new. Perhaps major labels and artists will get on board and produce something in a similar vein; however, this is not the only focus of Resound Worship as a network, and the grassroots level that was so crucial to the opening stages of the project remains part of the project leaders' overall ambitions:

I think what we were hoping to do was to try and make songs on this theme a conceivable genre for contemporary worship writers. Our broader mission as an organization is to resource local church songwriters and encourage local church song-writing. And while there will only ever be a small portion of songs that get widely published, I think there's profound value in having local songwriters engaging with this theme, producing songs on this theme. (Chris, interview with the author, 11 September 2020)

And so it comes full circle back to the songs contributed by different songwriters to the project. Ultimately these remain part of the mission, even if they were somewhat neglected in producing a marketable product. The relationship between market and community is a complex one, and while the demands of the market took precedence in achieving success, the market is not the final goal, and is used as a tool to achieve a broader influence and impact.

Beyond just worship

Resound Worship are not the only worship music group to have produced an environment-focused album in response to the climate crisis. A year or so after the *Doxecology* album came out, a USA-based collective, The Porter's Gate, released their own album project entitled *Climate Vigil Songs*, an album that, according to the accompanying worship guide, 'challeng[es] us to respond to climate change as an act of worship' (Climate Vigil, n.d., p. 1). The album project exhibits a number of similarities to the Resound Worship project but also comes with important differences.

In common with the Resound Worship project, the album features numerous different songwriters and artists over its 14 tracks, the result of sustained engagement and collaborations within The Porter's Gate collective, a shifting body of musicians, writers, theologians and artists who, over the course of the years, have produced albums focusing on themes such as work, justice, lament and being a neighbour. As with Resound Worship, the group sits on the fringes of the mainstream worship industry, attempting, in its own way, to build a dynamic and address themes that are not so often to be found in the CCLI top 100 listings. In a podcast interview with Joel Payne, Isaac Wardell, who convened the project, described the way in which he envisages the continuity between these earlier projects and the Climate Vigil album, with a sense that the driving theme of The Porter's Gate is one of hospitality, and that this hospitality entails not just hospitality to a range of different people, but hospitality to people's entire lives and the themes that concern them (The Resound Worship Songwriting Podcast, 2022). Reflecting upon the foundational principles behind the album, Isaac suggests that the album needed to perform a variety of different tasks and that this variety is crucial to producing something balanced and usable:

> Theologically speaking, we wanted this album to do three different things. First, we wanted to sing songs that celebrate the beauty and wonder of God's creation, with a particular emphasis on God as an actor in the world ... The second thing we had to do was pause and say, 'Something bad has happened here, and we are all responsible.' We need to feel sorrow for what has been lost ... Thirdly, we wanted to help people do something with these convictions, so we wrote songs for action and mobilization. (Climate Vigil, n.d., p. 5)

Rather than addressing itself particularly strongly to a single worship scenario, the album intentionally had a broader range of use-cases in mind. The songs traverse a wide range of themes, including a focus on God's mystery, creativity and hospitality; God's questions to Job about the creation of the world; a plea for mercy; a song written from the perspective of the Earth; songs of praise and of lament; prayers for the coming kingdom; a focus on jubilee, with its reward of the meek and punishment of the wicked; prayers for deliverance from dynamics of greed;

a song focusing on God's transformation of seeds and water; songs of waiting; and a song focused on consumerism and provision. In the podcast interview, Isaac suggested that some songs would be hymns for congregational singing, some would be more reflective and some would be call and response. The tendency for any one aspect to dominate was one that he intentionally avoided, steering away from a product that might be overly depressing in nature, that would be seen as simply a celebration of creation or that was completely focused on political action. Rather, he insisted that the album had to balance these different tendencies and create something that showcases a broader range of scenarios and possible responses.

Balancing different tensions

A focus on different use-cases means that this is an album in which different possibilities are deliberately set in tension with one another, precisely because of an awareness that certain avenues were likely to prove problematic in different ways. There is a tension, for example, between specificity and generality:

> What's always a struggle with these Porter's Gate albums is that they're always centred around a particular theme. And what that inherently means is that, as we start to write songs, there's this kind of built-in tension, that if you try to write songs that are really specific to the theological theme ... it can lack artistry and feel like what I call quote, unquote, message music. ... Or if you go to the other side and say, 'Well, we'll just make it general and vaguely hit at the topic', then it's easy for people not to even catch the theme at all. And so that's a tension that we always have ... to try to find that sweet spot of the songwriting process where it's clear that this song is addressing a particular theme, but it's doing it in such a way that it's just universal enough that it doesn't feel that it's too on the nose. (The Resound Worship Songwriting Podcast, 2022, p. 5)

This balancing act manifests itself in particular creative decisions and choices of words:

> Instead of just saying, the flowers, say the lily, or little things like that, that instead of being so broad, try to really get into the specifics of it. We wrote a song that I really loved. I wanted to write … a simple mealtime prayer song … The verses basically said, 'We thank you for the seed that went into the ground, we thank you for the rain that raised the seed, that brought it to our table', to try to connect ourselves to the Earth more than just being like, 'Oh, thank you for this fried chicken we have in front of us'. (Paul, interview with the author, 8 June 2022)

There is a careful balancing here of wanting to name things, of wanting to delve into particular themes and focus attentively on the world around us, while at the same time staying aware that sometimes this might work out and at other times might create songs that fail to speak broadly enough to be taken up or find an audience in particular contexts and settings. The album can push some boundaries, and play with different genres, but there are limits to this potential, in many cases self-imposed out of a desire to reach a particular public or audience. Just as with the Resound Worship album, issues of pragmatism form a particular point of tension, and one where it can sometimes be hard to strike a balance between expressing what needs to be expressed and a desire for that to be heard and received in particular ways:

> I do think we were a little too pragmatic. I think we could have pushed the boundaries more … People that aren't interested in the climate crisis … some little song that I write isn't going to change their mind … These songs, for me, were for people that are either working for justice or want to know what to do … You know, I think going in trying to write songs that change people's minds is kind of pointless. (Paul, interview with the author, 8 June 2022)

However, the diversity present on the album seems to have provided one route for the navigation of the different directions some songwriters felt pulled in:

> For me, it was such a hard album to write … It was a tough line trying to write songs that don't sound too preachy and finger wagging … versus the other side would be just like 'Oh, look at this beautiful world, wow, God has made a beautiful creation'. Those just sound boring to me … So I was trying to come at this crisis from a different angle. Obviously, we need songs of lament that are just God, what can we do, throw our hands up … but there's other songs … I don't know if we actually succeeded at this, but we thought it'd be fun to write a song that could work as a church song and also as a protest song that you could sing on the street. (Paul, interview with the author, 8 June 2022)

We can see some of the different tensions at work a bit more clearly if we turn to matters of emotion.

Acceptable emotions

The album brought together people from a number of different church traditions and, unlike the Resound Worship album, the norms of a typical worship set weren't something that always came so near to the foreground as a use-case for the songs:

> I come from a different context where I'm not doing six songs in a row … I don't feel like the church should be surprised when we sing a lament or when we sing a song, an angry song, you know, I think that should be a part of our prayer. We should have a well-balanced diet of joy, sorrow, anger, frustration … that's what our lives are like, so that's what our prayer should be like. (Paul, interview with the author, 8 June 2022)

To a certain extent, the diversity on the album extends into the realm of emotion, and this grappling with emotion was one of the key themes to come out of some of the interviews that I carried out. When I spoke with her, Kate outlined some of the different emotional perspectives that were drawn into the creative process:

> Grief, I think was one of them, but also wonder … I've been writing poems since I was five. And I feel like I kind of left off lyrical rhapsodizing about the wonders of nature. I haven't done that in a long time … Paul and … John Guerra had toyed with the idea of writing a song from the perspective of the Earth, how does the Earth feel? You know, how does the Earth show God's nature, God's handiwork, that kind of thing … 'Good Lord, deliver us' was frustrating in how often we went around on it and how much I just wanted to scream and not be pulled back. 'Declaring glory' was a much more positive experience because there's so much good that we could say … How do we narrow it down to which good things to say? … 'All creatures lament' was very much from a place of grief. But then things like 'Declaring glory', there was this wonder that I really had not attempted to tap into probably since I was a teenager. (Kate, interview with the author, 15 June 2022)

In a similar manner to the Resound Worship album, a variety of different approaches at early stages were gradually narrowed down into a product that was considered to work. Kate initially felt a sense of hopelessness and despair in relation to climate and environmental themes, and felt the need to channel this into her songwriting:

> Somewhere in there is where I decided that what I was doing was called rage hymning … They were very strongly worded laments … you get into the prophets talking about selling the poor for a pair of shoes … I did a lot of that, and it didn't go anywhere. I'm not sure in the current climate that that kind

of thing can go anywhere. Because you can't really work together with people who are screaming at you. So I wrote a lot
of angry stuff that got tossed by the wayside. (Kate, interview
with the author, 15 June 2022)

This anger was slowly moderated through a process of collaboration, and the desire to produce a more acceptable and
potentially useful product:

I still feel I was frustrated at a lot of times, because I felt like
... some of what we were writing ... didn't go far enough in
saying this is a problem and someone needs to fix it. When the
album came out ... I could hear that what they settled on ... is
probably a lot more constructive ... it's not alienating, it's not
demanding in the way that the very angry part of me wants to
be. It's more of an invitation for people to consider and make
decisions. (Kate, interview with the author, 15 June 2022)

This toning down is not, however, without a sense of ambiguity. More extreme emotions still feel necessary, even if an
album project such as this doesn't provide them with the most
appropriate home:

I do still feel pretty hopeless. It's 2022. And I have children,
and I look at the world ... What am I giving my kids? ... We
try to not be materialistic ... Reduce, reuse, recycle was the
big phrase in my childhood, and it's gotten us nowhere. And
I feel like the problem is so much bigger. And some of the key
players are so many layers above me that I as an individual
really can't touch it. But at the same time, I don't want to say
'Well, let's just give up, the world's on fire'. So yeah, I sit at
home and I write angry lyrics. And then I throw them away.
(Kate, interview with the author, 15 June 2022)

There is a collision between raw, immediate emotions and the
norms of acceptability that other individuals are ready to get on
board with, and this brings us back to one of the core tensions

at the heart of both album projects: the expectations of communities and institutions.

In their sociology of religious emotion, Ole Riis and Linda Woodhead draw attention to the ways in which different religious traditions shape the emotional lives of believers. They describe how religious emotional regimes offer a pattern for emotional life, and that this ordering offers a programme that 'clarifies which emotional notes must be sounded and which must not, which emotions should be foregrounded, and which should appear only on the "back stage" or not at all' (2010, pp. 76–7). It is not just the religious sphere of life that offers a particular patterning of emotions, however, and they suggest that the navigation of different emotional domains in the workplace, family or leisure activities has particular tensions in contemporary societies that can create a sense of fragmentation and confusion. They go on to propose that:

> There are also strategies for avoiding this. One such strategy privileges a certain emotional patterning as true and authentic, and merely 'acts out' required emotions in other emotional domains ... The experience of inhabiting varied and often incompatible emotional regimes leads to a heightened emotional reflexivity. This is likely to be greatest in the case of those who make the most stretching transitions across different domains ... Such people have no choice but to reflect on their feelings, and to become self-conscious about the ways in which they express them. (Riis and Woodhead, 2010, pp. 210–11)

The emotional ordering of worship often (rightly or wrongly) proscribes anger as an appropriate communal expression of emotion.[2] Because this is an emotion that Kate encounters and feels the need to express in response to other dimensions of the world around, she is sensitive to the current incompatibility of these different domains, and reflexively attempts to navigate the situation in which she finds herself. She does this, for example, through attention to broader questions of what a true picture of certain emotions really looks like. What does it mean to hope?

How does that play out in the current crises that surround us? How can those emotions be shaped according to a more realistic theological view of the world?

> I think it's really easy to talk about hope ... until you're actually confronted with these situations where it is hopeless ... There's an Emily Dickinson poem, hope is the thing with feathers that perches in the soul. Somebody wrote a response to it called hope is a sewer rat, Emily ... The way they describe it is hope as this tenacious, unkillable thing that even in the worst places survives and, goddamnit, it will drag you along with it ... I feel as Christians we have a lot of language of the hope is the thing with feathers, and it's bright, and it's sunny, and it's wonderful. And yet, here and now, I feel like hope is the sewer rat, and God is not cleaning the sewer. God is dragging us through it. But you know what we're getting through it, somehow. (Kate, interview with the author, 15 June 2022)

There is a great deal of tension here between different visions of hope and, in order to process them, Kate steps back from immediate issues of the climate to think about where different people's emotional landscapes come from, and how they are informed and shaped by different generational situations and expectations. The church as an institution has picked up on certain emotional narratives more than others, but if these are partly generational then there is the potential for them to be brought into question, just as a good understanding of God might have the power to question our own natural ways of relating to the world around us. We are caught up at the intersection of different world views and ways of being in the world and, while Kate seeks to mobilize reflection on these tensions in a healthy way, the intersection still continues to create tensions and dilemmas. Kate experiences similar tensions when it comes to matters of politics:

> Part of when I've written things that don't work it's not necessarily that they're bad lyrics or they're bad poetry. They're

not going to work in the context of congregational singing …
One of the songs that Paul and I wrote, 'Good Lord, deliver
us', we wrote that song on January 7th. So pretty much the
day after all these insurrectionists stormed the Capitol … It
started as this very angry political song that we rewrote a
couple of times … they tried to record it in Paris in September
and it didn't work … so in Nashville in February a big group
of us sat down and had to figure out how to rewrite this to be
more general, more about the climate, the state of the world
rather than a narrowly focused political take. And it was
hard … How do you say these things without just slipping
into spiritualese … just churchy jargon that just washes over
people that they can't hear any more because they're so used
to hearing it without using language that is so disorienting if
you hear it in a church, that people are just not going to get it?
(Kate, interview with the author, 15 June 2022)

As much as certain kinds of emotional expression, explicit
political position-taking is ruled out in order to produce an
acceptable product. There is a strong awareness that the normal
range of political expression might not do adequate justice to
the urgency of the situation, but also that the direct expression
of certain critiques and judgements feels too out of place to
work in this kind of forum. The eventual product is shorn of its
harder edges in order to become something that people feel able
to hear, but at the same time attempts to retain just enough of
them to have an impact and not simply leave people in the same
place as they started.

The limits and potential of the marketplace

In looking at The Porter's Gate and Resound Worship projects
we see both a microcosm of broader issues that different communities and individuals face in seeking to address the climate
crisis and a set of concerns that is unique and specific to the particular settings that they are focused on writing for. Questions

of diversity, expectation, traditions and change are common to a vast range of different settings. Perhaps more than any of the other projects encountered over the course of my field-work, however, these two albums are addressed to a context shaped by the market dynamics of popular music. The projects are intended to be downloaded, taken up, sold and utilized and they address themselves to a globally networked collection of communities and individuals who might potentially incorporate the albums into their own practices and liturgies. It is important that the music appeals to the right people and groups, and if it doesn't it is unlikely to do what it set out to achieve. The market is not the sole logic at work, of course; Christian communities – indeed any human community – operate according to multiple dynamics and organizing forces. But both albums are highly pragmatic in their desire to appeal to particular people and communities, and in both cases this leads to a certain funnelling down from an initially broad emotional or creative base into a product that can be taken up and incorporated into existing communal structures of worship and emotion without causing too much disruption along the way.

Climate-focused albums are not the first attempt to address socially pressing issues in the context of Christian popular music. John Lindenbaum (2013) argues that earlier attempts to include a social gospel dimension within Contemporary Christian Music exhibit neoliberal tendencies, including individualized lyrics, a communal address to the evangelical community as a bounded group, an avoidance of political discourse or state-based interventions, and a tendency to build on partnerships with Christian NGOs as the main collective avenue through which action can take place. There are obvious dangers to this approach, and Chad Seales's analysis of religion around Bono suggests that the intersection of music, evangelicalism, social concern and a neoliberal approach to the world presents us with an illusion, and that 'neoliberal religion as a cultural system keeps us from seeing other economic possibilities by making us think that there is no alternative' (Seales, 2019, p. 20). He suggests that, 'Bono is confident that he can love Africa, that he can transgress the colonial difference and still be the same' and that, 'Bono promotes the cultural practice of neoliberal religion as the secular consumption of a particular kind of spiritual affect – the feeling that personal choice has salvific consequences and that with this choice the consumer transcends the moral limits of traditional religion' (2019, p. 11). A consumption-based logic can hide the need for larger-scale change and give the feeling of doing something important while limiting the scope of what is accomplished.

The very fact that different groups and individuals engage in these forms, however, is a result of the latent potential within them. Mathew Guest has drawn attention to the potential power that is embedded within neoliberal logics, drawing attention to 'the cosmetic authority that mirrors styles and standards popularized in entertainment media ... a foregrounding of sentimentality as a dominant evangelical mood [and a] strategic authority that baptizes neoliberal economic virtues of enterprise, industriousness and instrumentalism for evangelical application' (2022, p. 491). While Lindenbaum's article shows what evangelical strategies deliberately avoid, Guest's demonstrates the reasons for their effectiveness – a power that is more than evident in the right-wing populism that has recently caused major upsets in US politics. We are faced with the challenge of figuring out whether there are helpful ways to harness this potential within the different limits that these forms and logics impose.

Andrew Mall makes visible some of the larger challenges that can be faced when trying to develop new directions in relation to established markets and communities. He suggests that progressive agendas can be complicated by a positioning in relation to broader Christian music industries, since these agendas can turn them into niche market phenomena, while market dynamics can result in the commodification of the very agenda that they are seeking to engage:

The challenge ... is to navigate a path forward that honors and furthers the values of its community without corrupting them, taking them for granted, or treating them as disposable ... Within music industries, ethics can strategically signify and bolster marketplace positioning and branding. This is true of individual musicians and bands who carefully craft their public personae. This is also true of organizations and companies, such as record labels and festivals, whose brands and reputations are significant among professionals (musicians and intermediaries) and consumers. The ethics of institutional and individual stakeholders in neoliberal capitalist markets are not necessarily diametrically opposed to those of Christian musicking, but neither are they wholly commensurate. (2021, pp. 108, 118–19)

It is a challenging space to navigate, and ready-made solutions are few and far between. Market dynamics mean catering to particular audiences, and that presents limits as well as opportunities. Neither of the projects discussed in this chapter seeks to radically reinvent their forms of worship or ritual; they both believe, to a certain extent, that it is important to work within the systems that are given so as to get people on side and ultimately achieve change in a pragmatic manner, even if they seek to introduce something new and different within that system. To a degree, the contributors and the project leaders are aware of these limitations. They know they're just trying something out, that it's just a starting point, and they hope that others will do something more. One reviewer of The Porter's Gate album lamented the lack of a more radical shake-up:

But these songs aren't a total fix for our anthropocentric music problem. Despite their ecologically astute lyrics, they remain the product of human music-making. They were written, performed, marketed, and purchased by humans. They speak of astronomical time, but their tempos are human ones, in the same metric ballpark as our heartbeats and footsteps. Their phrases are the perfect length for the average human lung capacity. These songs may conjure the idea of a musical creation that exceeds human song, but they do not embody

or enact that idea. Why is this a problem? Because the purpose of worship is not just to sing about justice and liberation
and renewal, but to invite them into being. Worship, as eco-
liturgical liberation theologian Cláudio Carvalhaes writes,
must 'start where it hurts.' It must respond materially to the
pain of the community in which it takes place. And at times
of crisis, this might mean radically disrupting our usual ways
of worshiping. (Parks, 2022)

Whether a more radical disruption is possible while also achieving the wide reach that these albums achieved is a question that
is hard to answer. As Ian Goodyer points out, there are ways
of engaging with the dynamics of the culture industry without
necessarily going along with the whole of its logic (2009, pp. 3,
137). It is possible to work with market-orientated cultural
forms without allowing the market logic to dictate the whole
dynamic of that engagement. It is possible to navigate these
dynamics, but the waters of doing so can throw up some dangers and challenges that need to be negotiated along the way.
On the one hand, it is possible to praise the kind of pragmatism
we find in producing the albums. If we want to have an influence
on the world, it's no good holding up static ideals that others
don't hold to and then wondering why they ignore what you're
doing. There is a price to pay in doing this, however. Certain
things will not make the light of day, certain emotions won't be
expressed, and perhaps some existing structures won't undergo
the kinds of challenge that might ultimately be required in a
fast-changing period of crisis and disruption. Later chapters
will describe some more radical alternatives. These each have
strengths and weaknesses of their own – they raise a range of
challenges to the activities described in this chapter, but this is
not a one-way challenge. It should never be the only option on
the table but, nevertheless, pragmatism deserves respect.

Notes

1 Not all initial responses were positive, and some expressed a range of Christian anxieties about the pitfalls of a particular focus or way of writing. However, it seems that at least some of the critiques that arose were understood as opportunities for constructive engagement. Reaching a broader theological audience, for example, meant encountering concerns and interest regarding the depiction of gender in the lyrics that might not be quite such a regular concern for an evangelical audience, concerns that Joel seemed to welcome and remain open to engaging with in his working process. Likewise, other individuals commented on what they perceived to be a narrow focus of particular songs, a concern that Joel understood but wanted to push back on, emphasizing the nature of the project as an album that balances different topics over a range of songs, rather than seeking to address everything in a single place.

2 See, for example, Wrenn 2021 or Blumenthal 2002 for arguments surrounding the potential of expressing anger in worship.

2

Activism and Acts of Protest

Singing, situations and movement(s)

In contrast to groups like Resound Worship, who aim to produce music in and for the Church, Christian Climate Action (CCA) are situated within the broader Extinction Rebellion protest movement in a way that foregrounds interaction with a range of different convictions and belief systems, creating music that both sustains the group themselves over the course of different events and actions and which is heard by other protestors and members of the public. Over the course of months and years, a whole range of different individuals within the group have put together different musical projects for different actions, experimenting, trying out different possibilities and addressing their creativity towards the different situations in which they find themselves or which they wish to create. Their activities have involved the production of a songbook for use during large-scale London protests, spontaneous singing during the same events, a climate carols songbook that rewrites Christmas carols with climate-focused lyrics, and a whole range of ad hoc events alongside.[1]

Music has become an important feature within the Extinction Rebellion protest movement more generally, and the music of the movement is massively diverse in character, performed by a broad range of different groups in a wide variety of different situations. As Donna Weston, Leah Coutts and Marcus Petz describe:

The breadth of musical activism within XR in the UK extends from a Baroque orchestra playing at the Daily Mail offices and a very large samba band tradition ... [Two XR activists] are establishing a 'Ukulele Army' to increase inclusive street musical activism, the instrument being accessible to untrained musicians. [There are] a wide range of XR related musical projects and arts groups throughout the UK, and 'because singing is so accessible quite a few XR choirs' ... Furthermore, songs have begun to be written down that are XR songs, available in written form as the Extinction Rebellion Noise/Song/Chant Sheet. Other chants and songs and even physical movements are spread in meetings both virtual, such as when using internet telephony, or in physical gatherings such as during occupations in Rebellion Week. Music in the form of chants plays a significant role in ecowarriors' lives in influencing activists. (Weston, Coutts and Petz, 2021, p. 13)

In the midst of this context, CCA projects carve out their own niche, both drawing on the wider Extinction Rebellion ethos and making their own distinct contribution to the dynamics of protest. Members of the group seek both to contribute to the wider movement as a whole and to carve out a distinctly faith-based way of being within it.

Music and ecological protest

Part of the reason for Extinction Rebellion's musical diversity is the fact that music serves a variety of different roles within protest situations. One popular framework for analysing the dynamics of protests describes these possibilities in terms of three principal tasks:

Through sharing their message and educating the public, many activists aim to create collective action frames (CAF), which are 'action-oriented sets of beliefs and meanings that inspire and legitimate the activities and campaigns of a social movement organization'. In doing so, activists may aim to educate the broader public, 'develop[ing] a message that tells others what the problem is and who is to blame (diagnostic framing), present[ing] the movement's solution (prognostic framing), and explain[ing] why the listener should join the movement to fix the problem (motivational framing)' ... Music in activism aligns with diagnostic and prognostic framing through its ability to be educational, and with motivational framing through providing a sense of community and belonging. [T]o recruit members for the new movement, activists must persuade aggrieved individuals that their problems are collective, urgent, and amenable to change. Music can play an important role here, as it enhances the urgency of action, shares stories and imagines alternate realities and futures. (Weston, Coutts and Petz, 2021, p. 3)

The collective action framing discourse helps to draw out some of the different tasks that are necessary within a protest movement, making it clear that we should not expect the actions described in this chapter to perform simply one particular role. However, as we will see over the course of the chapter, it stops short of doing justice to the full complexity that is present at the intersection of music, climate, religion and protest. The focus on particular tasks isn't always quite so strong as the analysis might suggest. Protest music can act prophetically, embody prayer, enable inter-religious solidarity, create dilemmas and unsettle bystanders, among a wide variety of other tasks, and assigning these to a particular framing category doesn't necessarily help to understand the interplay across different boundaries. Music can be deployed strategically or hopefully with a variety of modes of success and failure, and is highly situation sensitive, helping to negotiate particular relationships and individuality inside and beyond the group. It is precisely this mix of tasks that is behind the different activities that individuals try out throughout this chapter. For CCA, part of the complexity comes from the fact that protest is not something distinct and separate from worship, mission, prophecy or prayer, but rather is something bound up with these tasks in a variety of different ways.

The dynamics of protest

One of the first projects I came across at CCA was a simple one: the compilation of a songbook for use during a protest event in London. The book illustrates very well the situation-specific and interactive nature of CCA's musical activities but also the different tasks musical activity is called to perform in a situation very different from that of Chapter 1. It was compiled in the run-up to the Faith Bridge protest event in London, planned as part of larger-scale Extinction Rebellion action towards the end of 2019 as some members of CCA decided to put together a resource that would help the group to sing together during the protests (Christian Climate Action, n.d.[b]). The book incorporates a mixture of well-known Christian hymns, songs, spirituals and chants alongside Jewish, Buddhist, Islamic and other repertoire. It is not so much an attempt to create a new repertoire as to bring together existing songs that might be of use during the protests. Indeed, some songs have little direct bearing on climate issues but are, instead, included because they are well known enough to be easily sung together.

Charlie Wheeler, one of the main compilers of the songbook, suggested in conversation that he wanted to create something that would give the group artistic power and unite them as a movement. He found within this established repertoire that 'all of them speak quite well into that crisis and bring both a sense that we must be courageous in this time, and also that there is hope, but only if we're willing to act with integrity' (Charlie, interview with the author, 16 December 2019). While some in the group found the songbook helpful, a gap nevertheless quickly opened up between intentions in compilation and actual usage – some protesters found the song selection more helpful than others, while the practicalities of using a pre-prepared resource ended up having to negotiate the realities of police confiscations and spontaneous activity. It was precisely this co-incidence of musical activity and semi-spontaneous events during the protest that often led to interesting or moving moments. Charlie found particular power in the singing of 'Amazing grace':

So 'Amazing grace' ... it seems to be a way of non-violent de-escalation, the idea if you are singing and making a joyful noise, people's emotions are likely to be stilled a little bit ... at one point we were on one side of Lambeth Bridge and we were separated from XR south west ... we went on to the road and up through to Westminster Bridge ... and all the time we were singing ... it was quite a powerful experience because firstly there were loads of people from different faiths and we were all singing the same song, and what people experienced on the Westminster Bridge is suddenly these voices coming from over yonder ... it was some form of nostalgia to people ... being reminded that ... we're seeking peace, so ultimately we're seeking the good of the world, and so it's a reminder of beauty. (Charlie, interview with the author, 16 December 2019)

In contrast to many of the concerns about lyrical content in Chapter 1, here the lyrics of the song seem almost irrelevant – 'Amazing grace' has virtually nothing direct to say about climate change, and instead qualities of familiarity and nostalgia come together with a situation in which it became important to set a certain atmosphere. The music provides a means of managing the dynamics of the protest and setting a certain communal dynamic in motion (see Safran, 2019, p. 82). Protests are about the ability to affect different people, to create different impacts and to express something about the state of the world, and the music of faith serves a crucial role in guiding those dynamics in a healthy and productive direction.

Emotions and intersections

Different protests embody and interact with different sets of dynamics, and if we turn to some of the other activities that CCA have engaged in, we can see a variety of ways of approaching both protest and the role of music in enabling it to happen. A CCA protest outside the Church of England's General Synod meetings in February 2020 involved a mixture of communal singing and solo items performed by the singer-songwriter Samantha (Christian Climate Action, 2020). Since I was able to tune in to the event myself via the livestream, my own notes on the occasion summarize some of the event and how it played out from a distance:

As I tune in to a Facebook Live broadcast in early 2020, a group of Christian Climate Action activists are gathered on the pavement outside the Anglican headquarters, Church House, in London. Banners displaying the group logo and the slogan 'Children need us to act now' make clear both the identity of the group and the focus of the protest, while miniature models of coffins and flowers represent the deaths of children that have taken place around the world as a consequence of climate change. As cyclists, motor vehicles and members of the public pass the group, Helen Burnett leads the gathered individuals in a liturgy of healing and protest. Three musical performances are interspersed between prayers, liturgy and the reading of stories from children affected by climate change. The first, the well-known hymn 'The day thou gavest', is sung by the group led by Samantha on vocals; the second, a call for God's mercy in the face of humanity's wounding of the planet, is accompanied by a ritual action in which those present are invited to place snowdrops around the model coffins; and the third, a 'canticle of turning', celebrates the potential that 'the dawn draws near and the world is about to turn'. It brings with it grounds for hope and the possibility of a new future just around the corner.

In a subsequent interview I was able to understand some of why the music had taken the shape that it had, and what the event had meant for Samantha in putting it together and in the moment of performance itself. The musical dynamics of this protest showcase Samantha's own particular way of drawing together themes of faith, climate, community and spirituality:

I feel we've got so much to lament about and yet we don't do it, we are stuck in mainstream church I think in a 'you're in control, God, you're glorious, God' ... I feel like we are disengaged in some of the most important social justice issues of our time ... So it was really just ... naming where we are ... and then claiming Jesus in that situation ... lamenting, letting it emotionally sit and then drawing on Jesus to ... guide us into just action. (Samantha, interview with the author, 3 March 2020)

Samantha herself sits at the intersection of a number of communities and faith traditions, and this shows through here with some more evangelical emphases combining with traditions of activism and social justice. The song provides an articulation of truth on an emotional level and, beyond that, both an appeal to God and a plea to be drawn into action. It provides a nexus where these different realities and relationships are expressed and negotiated. Speaking about another song, she suggests that:

It's not a Christian song, but ... I don't like to draw too many distinctions ... it's a worship song, but it's deeply rooted in justice, and that question – could the world be about to turn? It's almost like speaking it prophetically ... it just felt so integrated, so powerful, and ... it felt honest in the way of giving space for lament and recognizing the death, the damage, the reality of what is happening now, and then also that, that embodying that prophetic calling of the church to speak ... it felt really explosive actually, it felt like a ... radical and true use of liturgy, ritual, music ... it's about bringing ... this tuning fork outside of the church building and allowing people to resonate with that as well, because if people are searching, they won't necessarily look in church. (Samantha, interview with the author, 3 March 2020)

In contrast to music that is focused on a bounded group or environment, music here provides not a clear articulation of one particular reality but a means for mixing together realities, of crossing their boundaries, and hoping that through acts of musical sounding something can stir or touch people. Indeed, this is one of the functions of civic spaces in general: they are locations in which different groups can assemble and express or enjoy themselves, and where they can do so in a place where encounter with a wider spectrum of others is not only likely but is expected. Such public spaces rarely stabilize into completely fixed formations; indeed, they resist this stabilization as spaces are continually re-purposed, re-used and re-occupied by a variety of different assemblies, publics and interest groups.[2]

Civic space, as Samantha describes it, is a realm that has unique possibilities for faithful action and true musical performance. Indeed, the possibilities that it offers, as described here, might even be considered more faithful and more true than spaces more explicitly dedicated to the performance of worship. It becomes true, I suggest, through its disruptive ability to voice stories and realities outside of the conventional limits of liturgical propriety, to bring Christian ritual into more immediate contact with the structural and personal realities of the world around, and to integrate a life of faith and a life lived out away from church buildings and the space for prayer and worship that they provide. It is not necessarily that such things are impossible in a church service, but rather that the conventions, structures and priorities that set such a space up as distinct and separate are precisely those things that order the world so as to put the challenges, connections and experiences of daily lived reality in a subordinate role. This layering of different realities and aspects of life together can be an immensely powerful one, with their combined significance sometimes taking on a greater significance than either would be able to achieve alone.

Beyond the boundaries of faith

Within situations of protest, dynamics of encounter are just as important as managing the internal dynamics of a group, and music serves a crucial role in setting the tone of these different interactions. While some encounters are the result of crossing paths with members of the public or with those policing the protests, others occur within the broader set of faith-based groups that are present at different events. As Kate Galloway has observed (2019, p. 4), climate-related activities can often provoke interaction across traditional religious boundaries. This ended up reflected in some of the musical activity that took place during the London protests:

> At one point we had a Muslim call to prayer, I found that particularly powerful ... I just found the prayer ... quite a powerful statement of ... solidarity with ... the rest of the Muslim world as well as the Christian world. (Charlie, interview with the author, 16 December 2019)

> If there's an aesthetic in music, you know there's a beauty in something then that's a great thing to have in a situation of possible ugliness, you know there's a beauty vs ugly power and police uniforms ... the music is always coming out of some tradition and mixing up with others ... it takes us out of our comfort zone a little bit, it should do all this, music, shouldn't it, it should be a little prophetic. (Mark, interview with the author, 9 January 2020)

Music, here, provides the vehicle for felt-commonality across confessional boundaries, offering a medium where sonic solidarity takes precedence over concerns of identity or articulated doctrine. Members of different faiths, alike, have an interest in the sacred and its relation to the environment, and music provides a means for experiencing some of this shared concern.

A project set around Christmas took a different angle in focusing squarely on encounters with the wider public, and utilized the medium of music in a way that drew upon the aspects of Christian faith they were most likely to be familiar with while subverting them to catch them off-guard and attract their attention. Towards the end of 2019, Grace Thomas decided to rewrite the lyrics of a number of well-known Christmas carols with climate-crisis-focused lyrics for a protest event outside a shopping centre:

> What I had in my mind when I was writing them, was that ... we would look like normal carol singers, and we would sing these ... sentimental, mushy kind of songs that everybody associates with Christmas. But then, when you hear the words, it kind of jars, and you take notice. (Grace, interview with the author, 3 December 2019)

Music, here, provides a strategic medium for play, for catching attention and for hooking into people's feelings. The lyrics often rework existing references to Christian imagery around climate-focused themes (see also Safran, 2019, p. 92). 'Hark the herald angels sing', for example, is rewritten beginning 'Hark the herald rebels sing! Let the sound of protest ring. East to West, and every child, Old and young are reconciled' (Christian Climate Action, n.d.[a]). This strategic deployment of Christian traditions is not purely pragmatic in nature, however, and it connects at the same time to particular visions of the Christian faith. For Grace, it is not primarily a theology of creation and the natural world that serves to drive her action and song-rewriting but rather a more human-focused concern for injustice and the effects of climate change on communities without the resources to easily weather the changes that it brings:

> The kind of theology that speaks to me is you know the liberation theology, the theology of the voices that aren't heard, and ... a lot of my climate activism actually is about wanting to raise the voice of those that aren't heard. (Grace, interview with the author, 3 December 2019)

The carols were initially deployed in an event outside a shopping centre; however, they also took on a broader life of their own. The impact and meaning of these carols is highly context-dependent, and Mark Coleman, who used them as part of a church-based evening of reflection, focused in particular on the beauty of the church event that they put on: 'the beauty, I guess, is just not feeling that we care together, we weep together, we laugh together, we hope together ... I suppose I'm talking about a beauty that is not isolation, and is worshipping a god that is hopeful and transforms all that is weakness into something potentially good' (Mark, interview with the author, 9 January 2020). Mark focuses on the reframing that needs to take place in order to find a form of Christianity that places not just ideas of other humans and social justice at the centre, but also 'a cosmic oneness sort of thing', placing the environment not as an added extra but as part of an integrated whole.

Symbols, invitation and pilgrimage

Part of the fascination, for me, of the musical elements in CCA events is their continual ability to emerge in new and unexpected ways as different individuals feel inspired to try something out. Protest has constraints, but these seem more open to a range of different forms and possibilities than the worshipping communities cited in Chapter 1, and this leads to a great deal of grassroots creativity. Barbara Keal's involvement in the pilgrimage to COP26 in Glasgow, 2021, can perhaps be understood as a kind of social performance art. In preparation for the pilgrimage, Barbara crafted a coat and wrote an accompanying song that could be used alongside the coat as part of the pilgrimage journey. The combination of coat and music together act as an important symbol for the pilgrims themselves and as a public invitation to interaction over the course of the journey. According to the website for the project:

> The Coat of Hopes is a patchwork pilgrim coat, made, worn and walked by hundreds of people over hundreds of miles from Newhaven on the south coast of England to the gates of COP 26, the UN climate summit, in Glasgow autumn 2021. The coat was made up of pieces of blanket into which were sewn or otherwise marked, the griefs, remembrances, prayers and hopes of people along its route and beyond.

The invitation to put on the coat is to experience in its wearing the warmth (of the love it has been made with), empowering you to carry the weight of responsibility we all share. The responsibility to respond with our whole self to the climate and ecological emergency we face together. (Coat of Hopes, n.d.)

Barbara describes the story of the coat, and the song that accompanied it, as follows:

I suppose the vision for the coat was about creating a piece of work that would connect people and would enable people to step into the sense of taking their personal responsibility and the climate breakdown story very seriously. But also has a kind of gentleness. And I think you can get the feeling it's got a sort of slightly legendary or historic kind of energy to it. So, when I was at the beginning of thinking of the coat, it felt really important to me that the coat should have a song. And I pondered long on what that song should be ... I think it has quite a folky kind of feel to it, but I think it feels kind of outside of a timeframe ... I studied sculpture at Wimbledon School of Art ... And really, from the middle of my first year, I had a specific piece of work that drew me into this way of working ... I wanted to find out if I could make an object that was powerful, but not through being crafted ... I started walking with a pilgrim staff to and from art school every day, which was about nine miles each way ... and I did that for two years. And in the process of that, the centre of my work moved away from object making, in a sense, and more into my own person ... What that walking work became most about was the encounters I was having with people in the street. So people would say, 'Where are you going?' as they'd see me striding through London with this staff, and I would explain that I was committing my walking to college as a pilgrimage by walking with this staff, using the staff as a symbol to help me anchor that commitment. And I would also ask them where they were going. So that you hear the echo of that in the coat song ... my concern as an artist ... was mainly around how we interrupt people within their everyday life in a way that causes them reflection or brings them into the present in a different way. (Barbara, interview with the author, 6 January 2022)

The song itself has two parts. The first is a repeated chorus inviting people to pose the questions of where the coat is going, its purpose and its name. These repeated choruses are interspersed with longer verses explaining the journey and the symbolism of the coat. As with many of the examples in this chapter, the relation to a particular faith is quite fluid. Barbara didn't want to frame the coat as belonging to any particular faith or group, but nevertheless felt that a kind of Christian energy fed into the creative process, and that on a personal level she's trying to do God's work with the project. A number of individuals suggested that the coat felt quite Franciscan in character while Barbara also suggested that the coat can be understood as a promise that we all belong together as we face the universal challenge of the climate crisis.

A sense of openness and invitation was crucial to the role of the coat in the pilgrimage, and Barbara described in interview how the invitation to put on the coat served as an invitation for people to stand in the role of protagonist in the climate breakdown story, and an invitation to stand in communion with one another, saying to the people making decisions that we're all here together, and that their decisions are being made for all of us, rather than simply standing and watching in critique. In interview, Barbara described some of the encounters that this invitation opened up:

We met these sort of 12- or 13-year-old boys coming out, they were on their way home from school ... and they saw us and I told them about the coat. And I said, 'Would any of you like to put the coat on?' And one of them says, 'Nah, nah,' and the other says, 'Nah, nah,' but the third one says, 'Yeah, all right.' Then, once he's put it on, the other two were just itching to get their turn to put it on as well … And it seemed to really open them up. They were excited by the idea of being part of history … and it sort of opened them up to wanting to talk about themselves.

And then outside the blue zone, where all the sort of important decision-making people were going … we'd … get there at about half-past nine. And then lots of delegates would all sing the coat song … And then we'd start to invite people to put the coat on … and were really moved in different ways, some of them weeping, some of them dancing, some of them saying that they experienced the coat as the only welcome they had had to this place … And then, interestingly, some few people who felt very clear that they didn't want to put the coat on … They weren't quite sure what would happen if you put this coat on. And they weren't willing to take that risk. (Barbara, interview with the author, 6 January 2022)

These experiences were clearly very moving for at least some of the people who were willing to try it out. There is potential for the invitation to change their attitude and change how they see their role, offering the ability to try on a different perspective and a different relationship to the scenario playing out around them. Alongside the experiences of trying on the coat itself, Barbara described the ritual of singing the song as helping to make sense of being in Glasgow after the pilgrimage had become a little more dispersed. She was sceptical that much of the current Extinction Rebellion repertoire had the kind of energy that the pilgrimage really needed, with a much softer tone than traditional protest repertoire, so the ability to fill this gap through other means and to offer the kind of sustenance that was needed was something she had to bring in through other means, with the song playing a crucial role in enabling that to happen.

Protest, strategy and the work of the Spirit

Relatively late on in the course of my research, one more protest caught my attention as a stream from the annual general meeting of Shell showed a small group of women singing a rewritten version of 'Amazing grace' in the middle of the meeting room. Once again, music had made its way to the centre of a protest, and once again the event had its own particular situation and dynamic. Kate, one of the protestors involved, explained to me the story behind the action:

> So there is an opportunity to do action in Shell and do some people in Christian Climate Action want to do it? You bet we do. So the question then becomes what is our contribution to this going to be? And one of the questions for us always ... is what can we bring to this that is uniquely Christian? ... And we're wanting to do it in a way that creates a dilemma for Shell, so that to leave us is disturbing, to remove us is awful and will not reflect well on them ... And we started thinking in terms of an act of worship ... And so the question was how to engage in an act of worship that was basically not going to be sacrilegious ... And so praying was a no-brainer ... but there was still the need to create more of a disruption and a disturbance. And so inevitably the mind then goes towards singing ... The question then became what do we sing and what's meaningful to sing in that situation. And 'Amazing grace' was alighted on quite quickly because of its symbolic links with breaking from colonialism ... And the other thing about 'Amazing grace' is it doesn't just belong to Christian culture, but it's a song that's understood by everybody ... Now, where it got a bit tongue in cheek, because obviously we adapted some of the words. And so what we tried to achieve was a blend between the uncompromising desire that Shell's activities and destruction should stop, but also an expression of the grace that the song itself talks about, and that we ourselves believe in, because what one should really be and ultimately hopeful for Shell, and for its directors, and

for everybody who's involved in it, is that they might see the world in a way that makes them want to stop doing what they're doing. And that's what we all need, we all need that chance to be given grace, and that the most destructive of us ... should be able to access forgiveness ... And so that's why one of the verses started, 'I pray God's blessing on those here that you'll become his heirs'. We're not here singing and protesting and wanting you to fall because we actually want your personal harm, we want you, along with the rest of us, to be able to move to a better place and be different people and be motivated by different things. (Kate, interview with the author, 7 June 2022)

Against a highly polarized backdrop of oil companies profiteering from climate-destroying extractive practices on the one hand, and protestors seeking to bring an end to these activities on the other, Kate's explicit framing of the protest in terms of worship and prayer and grace is a striking one. The action is clearly very strategic and aims to create a disruptive impact in a similar manner to other actions on the part of Extinction Rebellion more widely, but the whole thing is infused with a spirituality that may not be immediately visible to the casual observer. There is hope here not just to create a disruptive protest, but for the work of God in the hearts and minds of the directors on a very personal level, and to this extent it is a thoroughly spiritual act that is infused both by the intentions and inner dispositions of the protestors themselves, and by a hope and belief in the work of the Spirit in and through their actions. In the face of such hopes and aspirations, it is immediately tempting to pose the question as to whether these kinds of expectations are in any way realistic, and this is a question that Kate approaches from a range of different angles:

It was genuine prayer ... I genuinely wanted God to bless them. We were also making it very difficult for them to know how to respond to us. We were also finding a mechanism by which we stayed non-violent ... Do I think that the immediate effect on them would have been melting them to ... no. But I think this is where we move away from music to the purely spiritual. I believe there is something incredibly powerful about three Christian women walking into a room with, as I believe, the Spirit of God accompanying them. And our song and our prayer, it was a protest, and I'll be quite blunt about it ... I liked singing it at them, I'm very angry with Shell. But you have to temper that with love and the desire that they should be forgiven, not just punished. And I believe that we three women singing that in that place was an incredibly powerful thing, the consequences of which we will not on this side of heaven ... I think they'd all made the decision beforehand that if anything did kick off, they were all just going to sit there completely impassive. And that's what they did ... I have no idea what was going on in those individuals ... Clearly, a lot of the room were really cheesed off ... But others were ... on side with it because they were sympathetic to what we were trying to do ... one of the protesters in front of me was just welling up. So, for her, it clearly meant a lot more than just we were singing and it was noisy ... And then when we stopped singing, we were praying for [the policemen]. And I think that one or two of them were quite irritated by it. One of them, I think, was very moved by it. (Kate, interview with the author, 7 June 2022)

While the protestors hope that their actions might have deep and long-term consequences as those in the room come to repent of their actions and come to see the world through a different perspective, the immediate gesture of praying and singing is powerful in the moment, having an effect both on the protestors themselves and on those in the room. It is something that people see and react to, some positively and some negatively. Whether or not it has an outworking in the spiritual realm remains something unknown but hoped for, but the deeper spiritual and the more immediate practical levels are not easily separated from one another. Kate does not divide the world up into different realms but, rather, understands us each to face a basic choice to engage with the kingdom of God, or the opposite. This choice is continually at play in different ways in our daily lives, and is thoroughly integrated with the different actions we engage in in relation to the wider world around us:

> Engaging and connecting with the kingdom of God is to engage with it with that which is productive, which is enhancing, which fosters growth, is generative, is lovely, is joyful, is pure, is holy, is loving, is mutually supportive. And anything that creates that is connecting and tapping into the spiritual kingdom of God, whether or not the people involved actually recognize that. (Kate, interview with the author, 7 June 2022)

It is this broader world view that helps us to understand the role of music within the hopes and aspirations for the protest as a whole. Music plays a crucial role in connecting people with one another and doing something relational, a connection that Kate envisages in spiritual terms as something very much coming from God. It is, therefore, part of this wider sense of the kingdom. Through music's positive qualities, it serves to build something of the kingdom in different ways and different contexts – not just when connected explicitly with prayer and worship but because of what it is. The use of music as part of the protest is not simply oppositional in character; rather, it is a productive element that helps to tap into relational and spiritual dimensions of being and which, in doing so, might be able to help bring about a more positive future.

Effective strategies for musical protest

The diversity of examples and strategies on offer quickly prompts the question of evaluation. What does effective musical activism look like, and what can we learn from surveying a diversity of creative engagement? Through fieldwork conducted among various musicians in the area surrounding the Salish Sea, Mark

Pedelty set out to learn from experienced environmental activists what the key ingredients are in effective musical ecological activism. In the course of his journey he talked to children's entertainers, a collective of raging grannies, solo musicians and others, all of whom take a different approach to engaging ecological issues through music. Pedelty suggests that the environmental movement faces particular challenges when it comes to successful environmental musicianship. The movement is younger than established protest movements; its themes are harder to fit into established genres; many of its problems are invisible and hard to point a finger at; blame and hope are harder to apportion; and the movement is highly decentralized. While there are significant challenges to be faced, he suggests particular criteria for evaluating successful environmental musicianship. For Pedelty, successful environmental musicians are those who '(1) attract audiences, (2) work with environmental movements, (3) have some staying power, and (4) have effectively advocated for biodiversity, healthy ecosystems, and/or environmental justice' (2016, p. 17). However, he also observes how difficult it can be to draw a direct line between musical activism and policy impact.

This leads Pedelty to ask what exactly might enable this kind of success. After an extensive series of conversations and interviews, he lands on five themes that can be traced through the different groups he has surveyed and which, in some way, can be understood as contributing to their success (2016, pp. 237–54). The first of these is voice; Pedelty describes voice both as a metaphor for empowerment and as a visceral phenomenon, highlighting both the importance of gaining a voice in the public arena and the role that music can have in enabling this for communities. Place is about the deep connection between music and locality, music that doesn't simply use place to enhance itself but which is deeply embodied and carried out for the sake of a particular place and its future. Scale is to do with the relationship between different scales of action, and the relationship between local and global issues. Identity is about the need for a musician to be perceived as legitimate in the eyes of the audience. Humour is important in opening up the possibility for conversation that can often be closed down by more apocalyptic rhetorical strategies. Finally, communication highlights the fact that music is a crucial medium for communicative activity, communicating information, asking audiences to care, and doing so in a manner that is more than simply rational in nature.

While some of these themes can be traced across the different CCA actions, there are key differences between many of the groups Pedelty describes and the musical activities of CCA – in particular, there is less of a sense of performance and much more a feeling of group activity that is sometimes addressed to the outside. We can perhaps trace a slightly different set of themes and ingredients through the experiences described in this chapter. Diversity, situation-orientated creativity, the management of emotions and dynamics, reaching beyond the group and the inclusion of a spiritual dimension are all ingredients that come to the fore at different points. Each musical action is addressed to a particular situation or scenario that has commonalities with other actions but has its own distinctive goals and audiences which shape the form of action that can be taken. Likewise, each act of musical activity is tied up closely with the individual background, story and convictions of the person instituting it; these can be shared across the group more broadly, but clearly embody to a certain extent the characteristics of the individual creative actors.

One of the most interesting parallels to be drawn between CCA and Pedelty's ecological musicians is in the matter of humour. The groups do use humour strategically at appropriate moments but it is not their primary tool, and I want to suggest that there is perhaps a parallel to be drawn between Pedelty's discussion of humour and the inclusion of spirituality in the actions of the Christian protestors. If humour is a secret ingredient that serves to make a performance palatable and attractive, then a human desire for spiritual engagement might be another route in. This is not to say that this dimension will have a universal appeal, but rather that it is a dimension that can serve to curb some of the bleak reality of climate issues taken on their own.

The goals of activism

While Pedelty can evaluate success in terms of long-term audience and environmental engagement, the effectiveness of the CCA musical protests is not quite so easy to judge. If the goal is to reach the public, then reactions to Extinction Rebellion more broadly have been mixed, and CCA have achieved at most sporadic moments of wider visibility. Extinction Rebellion has served a crucial role in making climate issues present in public consciousness, and CCA occasionally make headlines when a vicar is arrested or taken to court for particular actions. But the group remains relatively small, and many actions fly under the radar for most of the wider population. If the goal is to reach politicians, then Extinction Rebellion have helped to build momentum around net zero targets and declarations of a climate emergency. However, the UK government is not currently on a particularly positive path with regards to climate issues and shows few signs of changing course any time soon, regardless of the involvement of Christian groups in addressing them. Indeed, they have hardened their stance against the actions of protestors and have sought to close down as many avenues for protest as possible, precisely as a symbol of their opposition to the different agendas that protest groups are bringing. If the goal is making an impact within the church scene, then perhaps the group can be seen as having greater success, making headlines in the *Church Times*, impacting debates in the Church of England synod and making ecological activism a possible and desirable choice for individuals and congregations across the country.

Music certainly plays a crucial role in this final alternative, helping individuals to feel a part of something, to integrate a spiritual and emotional dimension in their protest and to make themselves visible on, at the very least, a local scale as their singing can be heard by those around them. Creative responses allow a deeper integration of different dimensions, bringing spirituality and protest closer together than a reliance on existing options might always allow. In other research projects, I have attempted to think about the ways in which musical spaces outside of Christian worship often allow for a greater degree of freedom than regular communal services and allow the mediation of identity between different realms of life. The forms of climate activism that the group engage in are not always ones that are welcome in the context of Sunday worship. They go further than many congregation members are willing to go, or address issues that are believed to belong more appropriately in a setting other than those regular Sunday services. In the midst of this, many Christian activists seek out ways of reconciling different aspects of their identity together. As Maria Nita writes:

> Christian activists were often marginalized in their own faith communities due to their ecological beliefs or, at best, they felt that these beliefs were not represented in their home church and other religious organizations. Moreover, they were also a minority in the climate movement as both my qualitative and quantitative data illustrated ... The process of translating and thus integrating their Christian identity and their green identity may be seen as a way of allowing green Christians to operate from the very centre of their identity: 'transition in the name of Jesus' as one put it. (2016, pp. 159, 233)

CCA are not, perhaps, a group that should be understood as significant primarily in and for their own sake, but rather in their mediation between faith and activism, enabling Christians to inhabit both roles simultaneously. As we have seen, music plays a crucial role in enabling this to happen and is part of what sustains the movement in its identity while helping to reach out and build new connections and interactions. Its task is multifaceted and far from omnipresent, but key to the identity and energy of the movement of which it is a part.

Notes

1 This chapter should be read with an important caveat. The different projects described are far from pervasive, and they represent individual moments of creativity in response to a broader feeling that something is needed or lacking. Music crops up and appears all over the place but should perhaps not be considered the central feature of these protests; indeed, some individuals who I spoke with had almost forgotten that they had participated in any musical activities at all. Alongside the initiatives I describe, there is a broad range of musical activity, much of it less creative in nature, drawing upon established repertoire or simply using common Extinction Rebellion or protest chants that have been drawn in from elsewhere.

2 I am grateful to the editors of the *Oxford Handbook of Music and Christian Theology* for prompting me to reflect further upon these themes (Porter, forthcoming).

3

Communication and Song Festivals

Songs for an encyclical

While the protests of Christian Climate Action (CCA) are part of an outward-focused movement, aiming to provoke action and change on the part of government or corporate and ecclesiastical institutions that can rarely be assumed to be completely on board with their programme, Pope Francis' papal encyclical *Laudato Si'* addresses itself first and foremost (although far from exclusively) to members of its own institution. The encyclical aims less to provoke and confront, and more to motivate and inspire, prompting Catholics around the world to engage with environmental concerns as an important and theologically significant part of their faith and practice.[1] The discovery that a papal encyclical had inspired a range of musical projects was initially a surprising one to me. My brief ecumenical forays into Catholic congregations had not prepared me for the idea that official communications might provide a canvas for creative endeavour, and the existence of not just one song but several songs – and even song *festivals* focused around the document – was not something that I had been expecting.[2] What exactly was going on here? And did these projects offer something genuinely different from the other groups that I'd already engaged with? Or would they simply replay familiar themes, albeit in a context less familiar to me?

While there are numerous musical initiatives in some way inspired by *Laudato Si'* – an album of songs published by the Oregon Catholic Press (Canedo, 2017); a musical put on by a group of young people in Switzerland (Patrignani, 2022); and oratorios by the composers Peter Reulein (Băncilă and

Chertes, 2022) and Linda Chase (2022) – I was most intrigued by the existence of two livestreamed song festivals, 'Laudato Si' Festival "Songs for Creation"' (Vatican IHD, 2021) streamed for Laudato Si' Week and World Biodiversity Day in May 2021 and 'Laudato Si' Festival – Music event and talkshow from Assisi, on eco-spirituality' (Laudato Si' Movement, 2022) streamed in May 2022 from in front of a church in Assisi. I was intrigued by the ways these events integrated music into a wider institutional attempt at communication and outreach, and by the way they brought together a range of musicians from different contexts around the world, all of whom had taken a central message from the hierarchy in Rome and had chosen to build upon it in some way within their different individual contexts.

Two festivals, two weeks and a movement

Both the 2021 festival and the event in 2022 that followed it form part of wider movements and events centred on the encyclical *Laudato Si'*. In particular, while they were hosted by different Catholic bodies, both took place as part of a longer Laudato Si' Week, which itself was set up in order to celebrate the wider progress being made around the goals of the encyclical. During this week, Catholic groups around the world were urged to take up the challenge of *Laudato Si'* and to act together in caring for creation and meeting the environmental challenges ahead of us. According to the Laudato Si' Week website, the 2021 event 'encouraged the faithful to celebrate the great strides the global Church has made in its ecological conversion journey during the past six years and offered a clear road map for the decade ahead' (Laudato Si' Week, n.d.[b]). This included networking events where Catholics could come together and share some of what they'd been doing, participation in climate strikes, prayer and dialogue, and the launch of a new action platform offering a range of practical resources designed to empower Catholics to respond to the call to care for the environment.

The week itself is not an isolated event, however, and it is important to draw attention to the wider network of institutions and activity that has been put in place to implement the goals of the encyclical, particularly the Laudato Si' Movement (2021) and the Laudato Si' Action Platform (n.d.[a]), which was launched in 2021. The Laudato Si' Movement brings together a range of different Catholic organizations to focus on three strategic goals of ecological conversion, full sustainability and prophetic advocacy, offering resources, events, training and networking opportunities in order to mobilize the Catholic Church to implement the goals of *Laudato Si'*. A summary article by Cecilia Dall'Oglio (2020) highlights a range of different initiatives taken by the movement over the years, including petitions, campaigns and advocacy work, and the facilitation of divestment initiatives on the part of Catholic institutions. Dall'Oglio suggests that these programmes have made a significant impact, highlighting, for example, the influence of Catholic voices at the COP21 Climate Summit in Paris. The Laudato Si' Action Platform serves a similar role in connecting and motivating institutions, as well as linking them with appropriate resources to pursue concrete action and create strategic action plans. The platform focuses on seven key areas: response to the cry of the Earth; response to the cry of the poor; ecological economics; adoption of sustainable lifestyles; ecological education; ecological spirituality; and community resilience and empowerment (Laudato Si' Action Platform, n.d.[b]). Those who enrol, either as individuals or on behalf of different institutions, are encouraged to complete a survey that helps direct them to the most appropriate areas of the website. They are then encouraged to create their own Laudato Si' plan, which they can share on the website as a public commitment, inspiration and resource for other actors who can use it as a model for their own engagement and action.

Musical communication for social change

The role of the song festival within the wider Laudato Si' programme can be understood in different ways. The Laudato Si' Week website, for example, frames the festival primarily in terms of celebration, stating that the event in 2021 'brought together thousands to give glory to God through art' (Laudato Si' Week, n.d.[b]). But it also highlights the song festival's role as an act of communication, helping to share the importance of caring for biodiversity to a large audience of more than 8,000 spectators (Laudato Si' Week, n.d.[a]).

A variety of different frameworks can be deployed in understanding music that is designed to communicate, motivate and bring the members of an institution on board with a programme of action. Jennifer Publicover et al. (2018), for example, write about environmental musical interventions in terms of education and advocacy. In thinking through these categories, they draw attention to the role of music as a tool for communication, not simply of factual knowledge but of pro-environmental values. They draw attention to a range of artistic and creative decisions artists make in their attempts at environmental communication, positioning them on a series of scales between simple and complex, feel-good and not-feel-good, inclusive and adversarial, direct and ambiguous. If we follow this idea of communication further, then we see that a range of different models have been used when thinking about the role of communication in bringing about social change. Silvio Waisbord describes a number of different approaches that have fallen in and out of fashion over the course of the twentieth and twenty-first centuries. Waisbord differentiates between a dominant paradigm, in which problems of development are seen as the outworking of a lack of knowledge, and solutions are therefore orientated around the provision of information (2020, p. 95), and participatory theories (p. 108), which have questioned the basic premise of the dominant paradigm by drawing attention to a range of structural factors that contribute to developmental challenges beyond simply a lack of information. He describes a social marketing approach that can be seen to emerge around the 1960s, 'putting into practice standard techniques in commercial marketing to promote pro-social behavior' (2020, p. 99). Marketing techniques focus on the potential to influence behaviour and have, for example, been used in public information campaigns promoting healthy lifestyles or the use of seatbelts. This is not the only approach that organizations have taken, however. Entertainment education focuses around 'how entertainment media such as soap operas, songs, cartoons, comics, and theater can be used to transmit information that can result in pro-social behavior' (2020, p. 106), utilizing the natural draw of entertainment as a medium to simultaneously shift audience attitudes around particular themes and issues. Participatory approaches, meanwhile, include activities such as media advocacy, which aims to 'stimulate debate and promote responsible portrayals and coverage' (2020, p. 116) around particular important issues, and social mobilization, which involves a more 'comprehensive planning approach that emphasizes political coalition building and community action' (2020, p. 118). Rather than highlighting one particular strategy as the right way to go about communication for social change, Waisbord highlights the need for a toolkit conception of different strategies that can be deployed alongside one another in response to particular situations and needs as appropriate.

While these kinds of communication-focused approaches have been applied relatively rarely to music and the environment (Publicover et al., 2018, p. 2), Jessica Lansfield has done interesting work examining the use of rock music in environmental education through a social marketing lens. Lansfield focuses on a schools-based music project that aimed to use 'rock music and musicians to call upon students to change their consumption, question government decisions, affirm their environmental consciousness by taking action, and spread the pro-environmental messages and behaviours' (2015, p. 146). The project harnesses the relatability of rock musicians and the possibility of connection and interaction with them in order to motivate students towards environmental engagement and action. While Lansfield suggests that the students connected with the music and the message, she also suggests that better long-term integration and strategization were missing, so that the success of musical communication strategies will often be dependent on the opportunities available for engagement beyond the initial moment of encounter. Reviewing a short piece of video that was released in connection with *Laudato Si'*, Dominic Wilkins makes some similar observations, suggesting that Catholic social teaching offers a bigger view of the relationship between individual action and large-scale structures than a short-format video is able to include. Thus:

> ... while [the video] strives to avoid the disempowering emotions of fear and guilt by ending hopefully, its failure to provide concrete steps disempowers viewers by obscuring the severity of the problem and the scope of required changes. Without additional knowledge of Catholic teaching, [the video on its own] fails to effectively offer an alternative to contemporary hegemonic narratives of individual responsibility. (2020, p. 9)

How, then, can we understand the Laudato Si' Festivals as an act of communication? What is the role of music in a broader communicative ecology? Does the festival format successfully serve to mobilize a Catholic audience around a crucial issue? And how does it balance different paradigms of communication?

The programme of 'Songs for Creation'

Both of the song festivals that I observed follow a somewhat similar pattern. Over the course of a couple of hours, a presenter engages with a range of different guests, many of them priests or monastics, who talk about the importance of different issues relating to creation care and introduce particular projects or competitions, while every 5–10 minutes a different musician or group performs a song that they have written that focuses in some way on the encyclical *Laudato Si'*. A brief summary of the musical side of the evening serves to illustrate the range of different approaches on offer throughout the event.

The 2021 festival, entitled 'Songs for Creation', took place on a sunny evening towards the end of May. The festival lasted for just over one and a half hours and was streamed from an outdoor location, overlooking trees and sacred buildings in Rome. The evening is introduced and presented by Lucia Capuzzi, an Italian journalist, who is joined by a range of different conversation partners throughout the programme, alternating between discussion and pre-recorded songs and video clips. The opening song is from the Italian entertainer and project worker Luca Terrana. Luca's work includes raising funds for children living in shanty towns, and he uses music in order to bring people on board with the movement. His recording is somewhat repetitive, but highly memorable, featuring a mid-tempo soft rock accompaniment over which Luca repeatedly sings the phrase *laudato sii mi signore con tutte le tue creature* ('be praised, my Lord, through all your creatures') taken from St Francis's Canticle of the Sun. The second song comes from slightly farther afield, featuring an American musician, Steve Klaper, alongside a Franciscan friar by the name of Al Mascia. Steve is a Jewish cantor, and the introduction to the song focuses on the way in which people from different religions come together in harmony, something that ties into the two musicians' investment in longer-term inter-faith projects. The song itself is directly inspired by the text of *Laudato Si'* and features a more light-hearted offbeat rhythm, celebrating God's love for his creation and asking for his help in caring for the world and the poor. 'Africa vuka', the third song of the evening, is the first to feature a full band – in this case, a group from Kenya, filmed performing outside. The song is in a reggae style and features a repeated chorus calling on Africa to wake up, interspersed with rap emphasizing the need for systemic change, climate justice and a move away from fossil fuels and corporate profits for the sake of the next generation.

Music continues throughout the evening, and the next song features Gen Verde, an internationally networked women's performing arts group based in Italy. The song 'Turn around' was once again inspired by *Laudato Si'*, and focuses on a call for personal change. It is one of the more highly produced songs of the evening and begins by describing an encounter with an astronaut and the perspective that he is able to offer on the Earth when seeing it from outer space. The narrative emphasizes that the path to real change is difficult, but that with God's assistance it is possible to heal our common home. The song after this is the first to be filmed in a church environment and features the Argentinian group Filocalia singing a worship song celebrating the presence of God in creation, praising God for what he has made and asking for him to pour out his strength in order to help us care for our planet, for the poor and for those in distress. The song incorporates eschatological imagery of a celestial home and a joyful future. This is followed by a song from the Franciscan monk Sandesh Manuel, who lip-syncs to a recording of a song he has written entitled 'Listen to the wind'. The song asks how much longer we will live in a world where people are crying out, before emphasizing that we cannot carry on living how we are, suggesting that we need to listen to the wind, the silence and our hearts.

The penultimate singer, Migueli Marin from Spain, is filmed singing outside, with an optimistic message that he has peeked at the end of the Earth and seen an outpouring of beauty. He celebrates that this land is the greatest paradise we can have and features a catchy chorus calling on us to 'take care of her, of the generous Earth'. Of all the songs, this is the most upbeat, focusing not on problems and struggles but on the goodness and wonder of everything, and the responsibilities this entails. The final group, AAIRA, are introduced as rock stars who decided to make a change and perform Christian rock instead. The song once again begins with a celebration of the Earth before turning to the problems of human destruction and blindness to creation. In the face of this challenge it urges the listener to wake up from their indifference, to fight for the Earth and not to leave anyone behind.

On an initial glance, it is relatively easy to trace a number of common themes through the different musical performances of the evening – a celebration of creation, an identification of human failures and a call to practical action, often framed in terms of care. Some songs are possible to join in with as an act of worship, but many are framed as appeals, designed to turn the listener's attention to a particular way of seeing things, whether that be a celebration of creation or the need to turn around and do things differently. Many of the songs lift their language directly from the encyclical *Laudato Si'* and, to that extent, can be understood as a direct translation of the calls of the Pope into a musical medium.

From both musical and eco-theological standpoints, the songs present themselves as a mixed bag of sophisticated and simple, naive and knowledgeable, with an ethical appeal to the listening individual sometimes coming across as a slightly unreconstructed call to creation care, personally moving but naive to structural-level challenges and dysfunctions that we face.[3] Attention to the broader contexts out of which the songs arise makes it clear, however, that more is going on beyond the immediate impressions offered by the festival itself. When talking to some of the artists who had been invited to appear in the festival, it was clear that their songs had not been written explicitly for this festival. Rather, having engaged with the themes out of a variety of different motivations and in connection with different projects and groups, they had received an invitation to present their music at an event that seeks both to celebrate the range of work that has been going on and to motivate further engagement with these themes. Interviews with three of the musicians give a sense of some of this broader involvement.

Youth engagement and living the gospel in community

Nancy Uelmen is leader and mentor of the group Gen Verde, whose song 'Turn around' was included as part of the 2021 festival. Nancy described Gen Verde as an artistic expression of the Focolare Movement with the goal of building unity, putting the values of the gospel into practice in daily life and building bridges between people from different faiths, backgrounds and cultures. The group was born in the late 1960s as an all-female international performing arts group who, apart from the Covid-19 pandemic period, live and work together in community, putting on performances, concerts and workshops that, according to their website, 'give a voice to the people of our time, who are called to live for universal fraternity' (Gen Verde, n.d.). In interview, Nancy described her engagement with *Laudato Si'* and the origins of the song as something that is very much continuous with this broader project of engagement:

> We feel that our music needs to be an expression of our lives, so that the most important thing for us is to try and live the spirituality, which means living the values of the gospel, especially, which means trying to love each other ... and to build unity, through our daily actions ... We also have this project called Start Now where we do performing arts workshops and try and build this spirit through workshops with young people, and then they come on stage with us during the concert ... So it kind of came naturally to us to write songs about trying to live the values of the gospel in society, where there is injustice ... Obviously also the teachings of the Church are a big inspiration to us ... *Laudato Si'* was a big inspiration to me as I worked on that song. A lot of this is about translating some of those Catholic values into a public form that a lot of people can relate to. It's not about saying 'Gospel, Jesus, God', it's about finding the values that are common and sharing them. (Nancy, interview with the author, 17 June 2022)

The song itself had a double connection, relating both to themes important to the Catholic Church and to the values and interests of the young people present within the group and with whom they seek to engage:

Since we work a lot with young people, I felt that it was an issue that they really feel as important. So I thought we need to write something about this. And then getting into it, it was just like, 'Wow', in identifying with what young people are going through right now. And I also remembered an interview that I had done ... with astronaut James Buchli at the UN a long time ago. But he was already telling me about this experience of looking at the Earth from space and realizing how everything is deeply connected. So I thought, because often the songs we do, we try and connect them to our own lives [and] this is an actual conversation that I had that really touched me. And so ... I drew on that to kind of start the song. And obviously ... young people like Greta [Thunberg] and all the young people who are really in the frontlines are there. They were a big inspiration, but also the teachings and *Laudato Si'*. (Nancy, interview with the author, 17 June 2022)

As an act of identification, the song acts to build relationships, affirming the existing experiences of young people and putting that into dialogue with Nancy's own experience of dialogue in the conversation with James Buchli. However, Nancy's descriptions of the songwriting process highlight not just a generic identification with young people in general but a collaborative process of creativity that can speak to particular themes present within the Gen Verde group. In interview, Nancy described another song by the group entitled 'Uirapuru' on themes of deforestation. The song aims to express the experiences of two group members who grew up around the Amazon rainforest, using the image of a little bird to draw out themes of speaking up for what you believe in. 'Turn around' itself involves collaboration in the form of a virtual choir of children from around the world who were invited to participate via recordings and thus gain a voice and chance for expression as part of the performance. Through these kinds of collaborations, the group aims to affirm the experiences of young people on a range of issues, creating a participatory dynamic that enables them to speak out actively rather than merely as recipients of a message.

The work of the group does not consist simply in presenting songs in isolation, however, but often involves a broader programme of workshops that involve further participation:

> When we do the Start Now project, we're there for five days … We were in a city in Dortmund … and we went into the school and the kids all had their backs to us; it felt like they weren't interested at all in any workshops. And so the first hour, you could imagine it was kind of a challenge to get them to look up from their phones. But then, as we started doing the various workshops and stuff, they began to be very engaged. I think what helps them is that they feel that you believe in them, because this was like … a school for the kids that no one else would take. And they were really discouraged … We do a few days of workshop, then there's the day of the concert, the kids perform on stage with us. And then usually, if possible, the day after, we do like a couple of hours, like a feedback session … We finished that, walk out and go to the parking lot, and … two of them came up to us and this one of them had this clenched fist; he opens it and he says, 'This is 25 euros because we want you to continue doing this for other kids who are poor' … I think the strongest thing for them is to feel that we believe they can do it, and it kind of gives them an opportunity to build something different, to work together to do something bigger than themselves and maybe discover talents that they didn't realize they had. (Nancy, interview with the author, 17 June 2022)

At the time of the interview, the situation with the pandemic meant that the group had had relatively little opportunity to engage in workshops surrounding the song used in the festival, but Nancy's description of the opportunities they had been able to experiment with nevertheless suggests that the song could hook into a broad range of environmentally engaged projects and initiatives:

We did workshops with some young people here in Loppiano. And one of them was … a drama piece … which was also about the environment. And we did them back-to-back with the 'Turn around' song. That was online. And also, linked with the United World project – which is kind of a link to the youth of the Focolare Movement – there's this campaign Dare to Care … which does lots of activities to promote awareness for the environment, also local projects and stuff. (Nancy, interview with the author, 17 June 2022)

The song has use within a range of different settings, many of which emerge in somewhat ad hoc ways as part of different networks and relationships. It is a resource that becomes useful as different possibilities and opportunities arise to connect with groups and projects, and it exists as part of a broader toolbox that allows a certain flexibility to engage in different settings in ways appropriate to the people there.

Campaigns, awareness and a platform for peace and justice

Steeven's work is more activist and more integrated with a broader programme of ecological action than that of Nancy. While Gen Verde's involvement in *Laudato Si'* is connected to their broader ethos, it is far from the overall focus. Steeven, on the other hand, has recently started working for the Laudato Si' Movement, and has adopted the name Baba Miti, meaning father of trees, as a result of his broader environmental advocacy and tree-planting campaigns. His role as a lay Franciscan has offered training to work for justice and peace, as well as giving him a wider network with whom he can work on human rights, justice and the care of creation. Steeven's own experiences of environmental and human destruction have shaped his activism in important ways:

I always go back to the stories like during those times of genocide … I could remember that forest trees were cut down,
bridges were broken, so that nobody can escape … And I saw
wild animals running away, rabbits running away, having
nowhere to hide. People couldn't find where to hide, because
trees have been cut down, and the rivers turning red because
of the dead bodies that have been dumped in the rivers …
So, today, when I'm talking about preaching *Laudato Si'* I
can relate how peace and environment is connected. (Steeven,
interview with the author, 15 July 2022)

Having witnessed first-hand the environmentally destructive
effects of genocide, and the plight of both animals and humans
in the face of environmental destruction, Steeven sees a close
connection between environmental concern and issues of justice,
human rights and peace. Living through these situations
together with the society around him moved him to become
an activist. In this context, music became a means of bringing a new generation together, of using their talent, and a
way to bring a little joy to other people. He understands his
music to have a prophetic role, calling on the nation to change
and warning of the consequences if they don't. Within these
broader projects and involvements, the release of two songs
came as a response to the potential opportunities offered by the
broader programmes surrounding *Laudato Si'* and their arrival
in Africa:

When we started now celebrating Laudato Si' Week, the
season of creation, I said 'God, this is the right moment for me
to do what is right to preach for the environment' … Laudato
Si' … opened an African office in 2020. And there I said, 'OK,
now I want to contribute to this office to do awareness so
that in African countries they can embrace this encyclical
letter and this global agenda.' But when I launched this song,
'Laudato Si' celebration', I was surprised it was immediately
taken to the Vatican before even Africans knew the song. It
was already trending abroad … So that is how I realized that

maybe … let me do an African one now … That is how I decided to do the 'Africa vuka' … *Vuka* means like transition to close the border from fossil fuel to renewable energy. That is the key message in that song. (Steeven, interview with the author, 15 July 2022)

The second of these songs, 'Africa vuka', was included in the Laudato Si' Week celebration, and is explicitly addressed to political leaders and their failure to meet climate action goals and agreements. It incorporates Greta Thunberg's use of the phrase 'How dare you' alongside a call for Africa to wake up, and to prevent corruption, environmental destruction and the use of fossil fuels, investing instead in renewable energy. It brings out themes of climate justice in opposition to greed and, in this sense, it doesn't simply present an abstract and spiritualized view of ideal environmental relationships but seeks to mobilize a broader base of the African population to demand concrete action from their leaders.

Steeven is keenly aware of the strategic importance both of a church-based platform and of music as a medium. Each of these offers him the potential to get a message out when other avenues might more easily be ignored or shut down:

But now we are lucky. Laudato Si' is a noble platform … It is safer for me to go through church than to go through a secular world because in a secular world whatever we used to do … people think that we're doing politics or … we are doing it for money … But today, if I'm preaching this environmental justice, using the story of Eden Garden, people can understand that I'm quoting the Bible, I'm doing the right thing … There is also something unique in the music because … nobody can stop me … I don't need to ask for permission. I stand up, write my music, sing it, address the person I want to address … you can close my mouth, you can even eliminate me but that song has gone out, it will never stop its work … The second strength of music is that the music … it calls us, because people need to hear music even during funerals, people play music … you can call even somebody who did not eat to start smiling and dancing. So there is a power in the music in communicating. And … it brings people together … you will see people dancing as a community. So music will make people celebrate while they are also getting a message. (Steeven, interview with the author, 15 July 2022)

The song has played a role in helping Steeven to develop different relationships and connections, putting him in touch with other activists and platforms as well as different figures within the life of the church. He has the feeling that it has helped to enable his message to be heard by a wider audience, perhaps even an audience that would be impossible to reach by other means.

Relatable content, visible spirituality and life

While Nancy's and Steeven's projects both carry with them a broader attempt at social change, my final conversation partner, Sandesh, was a little more cautious in the framing of his project, suggesting in interview that:

I really care about creation, as I told you. This has been a very big thing for me. I believe I cannot change the whole world. That's my favourite quote, you change yourself and there will be one less idiot in this world. And the only thing I do is … doing such videos makes me aware of my connection to environment, my consciousness to and responsibility to environment. I'm not an environmentalist, I'm not an activist or something. But these are small little drops of ocean, which also convey the spirituality of St Francis. And make me also aware that I need to take care of a mother, Mother Earth. (Sandesh, interview with the author, 8 June 2022)

While Sandesh initially distances himself from any broader form of activism, and frames his own responsibility instead on a personal, individual and spiritual level, his work is by no means completely isolated from larger scales of engagement. His project, in producing songs and music videos for YouTube, is partly a personal one; he is engaging in something he enjoys, using his skills and talents, and taking them in a variety of different directions. However, it is also an attempt to put something out in the world that a wider audience has a chance to resonate with. His videos make visible a mode of Franciscan spirituality that embraces life, provides a source of meaning and enjoyment, and which has room for the different quirks and interests that Sandesh chooses in different ways to showcase. Sandesh's engagement with environmental themes comes from observing societal trends around him, but also through connecting these to a deeper sense of Franciscan spirituality:

> It was inspired by *Laudato Si'*. And then I started to write and think about this song. And within my mind also were the times when Greta Thunberg and the Fridays for Future were picking up ... and people, children were marching on Fridays here in parts of Vienna ... Being a Franciscan, it is automatic, we are born in the spirit of St Francis of Assisi, who was so close to nature. You might have heard of the film *Brother Sun Sister Moon* where St Francis calls all the creation as his brothers or sisters because he believes God is our father ... So it is not a new theme today, and Franciscans have always been giving their kind of potential for this ... so, being a YouTuber, I make many things like racism, like environment, the humanity piece, I want people to get together and all that ... When I was a student we used to do street plays on environment and all that and now I'm just doing music. (Sandesh, interview with the author, 8 June 2022)

Sandesh seizes on the opportunity afforded by both *Laudato Si'* and the Fridays for Future movement to showcase the connection between current events and the longer trends of Franciscan spirituality. It is an attempt at communication and connection, but one that is framed by a broader attempt to showcase his own Franciscan and personal identity in a public environment, allowing him to engage in creativity, produce content and build up a subscriber and supporter base interested in what he has to offer. Some of this seems intentionally designed to appeal to, inspire or communicate with a youth audience, showing them a particular way of being Franciscan or Catholic and drawing attention to a shared humanity rather than arising purely as an act of self-expression. In this sense, Sandesh's work is framed by the broader logic of YouTube and social media and also by the logic of entertainment. According to Sandesh's website, 'His passion is to see people smile … and that's why he is a Musician, painter and a YouTuber. Believes in Humanity and is open to everything what [sic] makes peace possible' (Manuel, n.d.). Sandesh is interested in breaking out of boxes and in subverting existing prejudices and preconceptions that people may have about what it means to be Catholic or a monk. He wants to show that he is human too, and that his interests are not all that different from other people's, but that Franciscan spirituality is something he has found valuable in navigating life (WDR, 2022).

As an expression of Franciscan spirituality, the original video for the song taps into a wider sense of Catholic community, featuring shots of both Sandesh and other Franciscan friars dancing in different outdoor locations, and to this extent Sandesh's personal platform provides a vehicle for a wider sense of mobilization behind a common cause. This mobilization of community is momentary rather than sustained but, through his wider YouTube presence, Sandesh's video manages to draw briefly to the foreground a possible resonance between Franciscan spirituality and current concerns relevant to his broader subscriber base. The video hooks into a longer-term project to craft a resonant Catholic and personal identity drawing together life and faith, a project that, from Sandesh's descriptions of people's reactions, may not appeal to everyone but certainly connects with some. The song communicates that there is no disconnect between care for the environment and life as a Catholic and, as such, serves as a form of integrative communication, creating an image of Catholic spirituality that resonates with current societal concerns rather than jarring with them.

Communication, connection and celebration

Surveying these different projects, we begin to see some of the broader context that the Laudato Si' Festival hooks into. The different songs that are presented each relate to a multi-dimensional communicative project of their own that is rooted and gains meaning outside of the festival context but is connected with both the festival and a range of other activities through the unifying force of the encyclical.

On a superficial level, the Laudato Si' Festivals can sometimes present themselves as over-simplified, often lacking in a specific call to the most helpful forms of action; they can sometimes focus slightly too strongly on the individual level, striking an overly upbeat tone; or they can reinforce a narrative of creation care that maintains the status quo rather than addressing the disruption that we are currently living through. As such, they can resonate with critiques of *Laudato Si'* that highlight its potential failure to transform moral awareness into broader programmes of structural-level change (Buckley, 2022; Jamieson, 2015). The repeated emphasis on the message and wording of *Laudato Si'* throughout the festival evening serves to highlight a common identity and common project, but also risks over-repetition of a single facet of a complex and deeper agenda.

A deeper exploration of the role of the festival and the musical artists as they connect to a broader programme of thought and action, however, suggests that such a critique may be overly hasty and that, for those involved in the broader programmes that *Laudato Si'* represents, the role of the festival as a celebration of that work, a place to come together and a creative space for networking and awareness, may well go deeper than a casual viewer is aware of. Indeed, this potential dissonance between impressions and reality highlights precisely the need for communicative acts to be embedded in a deeper ecology of action and engagement in order to become meaningful.

As a result of the broader integration of the encyclical into Catholic hierarchies and structures, the performers at Laudato Si' are able to tap into a strong sense of Catholic identity when drawing upon the text. While elements of *Laudato Si'* and the festival agenda can be understood in terms of a top-down or presentational communication model, there are important participatory dimensions to consider. The artists involved in the festival built up their different projects independently of any direct call for music to be written, and received invitations to participate in the festival as a response to those different endeavours. Likewise, the projects themselves involve important participatory dynamics, bringing a range of different actors on board for the performances, and using them as part of broader collaborative or interactive projects in the different locations where they are based. Environmental communication, when understood as communication for a particular purpose, is meaningful to the extent that it connects to behavioural changes and action on a wider scale. The connections between the festival and deeper programmes of action could have been more carefully brought to the fore during the course of the festival evening, but they are clearly present behind the scenes, and the variety of the different projects suggests that action is broad rather than narrowly focused on one particular avenue. The Laudato Si' Festival adds important dimensions to the projects described in other chapters precisely through its emphasis on the role of musical outreach and communication and through its integration of musical activity with wider institutional programmes of transformation and development.

Notes

1 The encyclical *Laudato Si'* (Francis, 2015) is a long and seemingly comprehensive document outlining the reality of climate change and biodiversity loss, their consequences on a social and planetary level, their moral and theological importance, the weak responses that have been undertaken up to this point, and the structural obstacles that lie behind current failures. It draws attention to spiritual and biblical wisdom on our relationship with creation, resisting an ontology of anthropocentric domination and pointing to the diversity of creatures and relationships that should be affirmed and celebrated. The encyclical highlights the

possibility for more productive dialogues on a range of levels and the importance of ecological education and spirituality. In doing so, the document contributes to Catholic social teaching on the environment, aiming both to educate and to motivate individual and structural action on environmental matters within the Catholic Church as a whole and on the part of anyone sympathetic to its aims. The authority of the document, and its ability to draw in the resources of a vast international hierarchy of sub-organizations and leaders, mean that, while it has been subjected to a number of different critiques, both theologically and in terms of its practical outworkings (see, e.g., Buckley, 2022; Flores, 2018; Grey, 2020; Heuvel, 2018; Jamieson, 2015; Nche, 2022; Wilkins, 2022), it has been perceived by many as a landmark document, and the encyclical has been celebrated by many both within and outside of the Catholic community.

2 On reflection, the number of musical projects inspired by *Laudato Si'* can likely be put down to a combination of different factors, combining the influence and reach of the Catholic hierarchy, its role as a document that is supposed to motivate and inspire a programme of institutional and individual action, and the resonance that many felt as the result of the Catholic Church taking such a prominent stance on an issue that they already felt was important. This combines with the document's roots in a canticle by St Francis, a title that can easily be translated into a song of praise, and a broad appeal that is not limited simply to worship contexts, to explain why so many have been inspired to write music in response to the encyclical's call.

3 To a certain extent, these tendencies seem to be fundamental to the nature of the encyclical itself, and existing literature on the reception of *Laudato Si'* suggests that, at least in some contexts, the encyclical has tended to operate more on a personal moral level than in motivating more structural-level engagement. David Buckley, for example, has suggested that 'consistent with recent research, evidence of a "Francis effect" is more robust in raising the moral salience of climate change, and less so on impacting policy attitudes' (2022, p. 4).

4

Grief and Ecological Requiems

It's welling up as a need, [and] I think the church is stepping up to meet that need, which I find very moving because, like climate change, grief is a genuine feeling that I think a lot of young people particularly are feeling – this deep sorrow, this deep grief. And because the church, you know, well, we have ways of dealing with grief ... People of faith – whatever faith – we've got a container. And so we've got the experience of creating a container for people to explore these uncomfortable, painful emotions. And I find that very exciting, actually, because it is also a way of offering hope. A lot of people involved in climate change activism are struggling to hold on to hope and it's something that people of faith are able to offer, I think. (Barbara, interview with the author, 8 October 2020)

Lament and sorrow

In recent years, it has become increasingly common to see the performance or composition of requiems focusing on the changing climate or on the numerous different species lost as a result of ecological destruction. Extinction Rebellion have framed a number of different protest events in terms of a requiem;[1] the sound artist Steve Norton recently put together a sound installation called 'Requiem: in memoriam of twelve recently extinct species' in collaboration with the Goethe Institute in Boston (Goethe-Institut, 2021); and Celia Damström composed a 'Requiem for our Earth' as part of the Ung Nordisk Music festival in Tampere (Damström, 2019). We can also find numerous more personal attempts at requiem-creation on

YouTube and SoundCloud, and we are beginning to see a range of commissions and performance events taking place within more established institutional settings.

The growing popularity of the requiem as a format is a result of an increasing recognition of the different extinctions and losses that are happening as a result of human-created ecological destruction and the need, in some way, to process these losses and what they mean. Within Christian circles as much as beyond them, a number of different voices have tried to draw attention to these dynamics, seeking to recognize the psychological processes many of us are already going through, our own culpability for the traumas of our planet, the denial that prevents us from processing these emotions properly, and the possibilities that there might be something positive to do before everything gets worse. In 2019, the ordinand, doctoral researcher and activist Hannah Malcom won the *Church Times* Theology Slam competition with her exhortation that 'all of us can experience a form of unnamed melancholy when places we love get destroyed ... This is solastalgia, and climate chaos will create unavoidable homesickness for all of us' (Hymns Ancient & Modern, 2019). The theme of climate grief has slowly emerged into the mainstream within Christian circles and beyond as it has become ever clearer how high the stakes are, how little action is being taken and how much we have already lost.

In this chapter, I want to reflect on conversations around two climate requiem events, with a particular focus on the creative process involved in putting them together. First, with Jonathan Arnold, Adrian Bawtree and Emma Pennington, who have been working together to produce a requiem for extinct species as part of their involvement in the life and work of Canterbury diocese and cathedral and, second, with Shirley Thompson and Roderick Williams, two composers involved in a commission for Christian Aid and the Chineke! Foundation recorded in mid-2021 in St Paul's Cathedral and broadcast via YouTube (Christian Aid, 2021b). I want to focus in on the wide variety of different influences, perspectives and concerns those involved draw together in imagining and staging these events, and to reflect on the positioning and role of the requiem format as an intersectional medium for creativity and performance. Through these conversations we will see some of the different possibilities available in navigating grief and the different strategies that can be used to shape this grief in a way that is helpful both for individuals and for the broader world around us.

Wider contexts

The question of how to mourn and grieve in worship has long been a theme for debate elsewhere in the Christian music world. There is, in particular, a rich sub-genre of evangelical blog posts bemoaning the lack of lament songs in Christian worship. One of the best-known early expressions of this critique came from the writer and pastor Brian McLaren in his open letter to worship songwriters, originally published in the early 2000s:

The Bible is full of songs that wail, the blues but even bluer, songs that feel the agonizing distance between what we hope for and what we have, what we could be and what we are, what we believe and what we see and feel. The honesty is disturbing, and the songs of lament don't always end with a happy Hallmark-Card-Precious-Moments cliché to try to fix the pain. Sometimes I think we're already a little too happy, excessively happy on a superficial level: the only way to become more truly and deeply happy is to become sadder, by feeling the pain of the chronically ill, the desperately poor, the mentally ill, the lonely, the aged and forgotten, the oppressed minority, the widow and orphan ... This pain must find its way into song, and these songs must find their way into our churches. The bitter will make the sweet all the sweeter; without the bitter, the sweet can become cloying, which is why too many of our churches feel, I think, like Candyland. Is it too much to ask that we be more honest? Since doubt is part of our lives, since pain and waiting and as-yet unresolved disappointment are part of our lives, can't these things be reflected in the songs of our communities? Doesn't endless singing about celebration lose its vitality (and even its credibility) if we don't also sing about the struggle? (2011)

Such sentiments have been repeated in countless blog posts and opinion articles and have led to the production of albums by groups such as The Porter's Gate (2020), Cardiphonia (2020), Bifrost Arts (2016) and others, mainly sitting somewhere towards the margins of mainstream worship music. More mainstream songwriters and bands have also ventured into this territory, but more often on a single-song basis rather than offering album-length treatment. At the same time, much of this has failed to make its way into a mainstream environment as solutions that need to be structural in nature are often addressed on a repertoire level. New songs are produced and recorded, but quickly get pushed to the margins of a congregation's repertoire because the forms of worship that it regularly employs have no easy home for songs that push too far beyond their standard affective and intersubjective patterns of engagement. These projects point to the need to find new ways of processing troubling emotions but don't yet offer all the answers, continually trying to do some justice to people's feelings but too set in other patterns to consistently move these to the centre.

It is not just in the Christian world that themes of grief are receiving increasing attention, however, and themes of grief, anxiety and emotion have become increasingly important social questions in light of the mounting ecological devastation we see around us and the increasingly visible toll that these emotions are having on different people and communities. In Ashlee Cunsolo and Karen Landman's edited collection *Mourning Nature* (2017), the different authors draw attention to the importance of grief for the world around us, while examining the ways that grief and mourning change when they move beyond the realm of the human. They describe how, on the one hand, it can be hard to know how to mourn non-humans, and even more of a challenge to mourn on the more abstract level of ecosystems and assemblages (2017, p. 16); and, on the other hand, how mourning can open up new relationships of empathy (Braun, 2017, p. 86), and can be used to expand discursive spaces around climate change (Cunsolo and Landman, 2017, p. 13). They suggest that analogies between mourning nature and mourning for human deaths can help the process (Menning, 2017, p. 40), as can a focus on specific beings and situations (Menning, 2017, p. 44). This focus can be matched to the specificity of a particular loss, but this can often tie into a broader landscape (Cunsolo and Landman, 2017, p. 18), with the loss of a particular species tied up with the loss of certain feelings, experiences, landscapes and a sense of place (Whale and Ginn, 2017, p. 100).

The authors suggest that nature mourning is often future-orientated, anticipating future loss not just of animals and ecosystems but of ourselves, humanity and the world that we live in (Braun, 2017, p. 83); and that while the goal of mourning can be highly pragmatic, focused around readjustment, it can also act as a restorative spiritual practice (Braun, 2017, p. 74), reawakening bonds and connections with what was lost (Cunsolo and Landman, 2017, p. 22). This mourning need not remain individualized but can become a powerful and political act. It extends the reach of kinship and recognizes vulnerability (Cunsolo and Landman, 2017, p. 14). Moreover, it can become 'resistant mourning' (Barr, 2017, p. 191) through a refusal of consolation and an acceptance of responsibility (Barr, 2017, p. 193). It can become a protest against unjust structures (Cunsolo and Landman, 2017, p. 14); it can serve as a mechanism for political mobilization and counteraction of dominant discourses (Cunsolo and Landman, 2017, p. 19); and it can foster ethical community (Cunsolo, 2017, p. 179), acting as a place of orientation and offering a new vision. There's a lot going on here, but as I talked to different composers, many of these issues came up as they sought to shape their different requiems and requiem movements in a way that would give people what they needed, not just to mourn but to act, to orientate themselves in the right direction for the future, and as they sought to move from the human-focused genre of the traditional requiem to a composition that addressed something traditionally beyond its bounds.

Concert Masses, civic spaces

All of this is possible because of a general shift in what Mass performances are able to do. The revival of the requiem as a reworking of traditional liturgical and musical formats provides a ready resource which can be reshaped in a way that provides a locus of meaning not just for Christian communities but for a wider public, hooking into a current need to process the destructive elements of human activity and the negative emotional landscape that goes with them. The requiem is a form that has acquired a degree of recognition and cultural capital beyond the bounds of Christian community; usage over a period of centuries, combined with treatments by high-profile composers that have entered into repertoire, mean that a performance of a requiem is able to be opened up not simply for a worshipping community but for all who are seeking a locus of meaning through aesthetic forms.

In her 2015 doctoral dissertation on the changing nature of musical Mass-setting, Stephanie Rocke dedicates an entire chapter to the theme of concert Masses for the environment. Throughout the thesis, Rocke argues that, as a result of religious diversification, the sacred Mass has gradually moved from its medieval ritual form towards becoming a politicized concert event, with 'social, economic and political responses to Enlightenment thinking also contribut[ing] to the devolution of the mass from its institutional origins to its unrestricted chameleonic form' (2015, p. 275). Rocke suggests that 'a form that was once wholeheartedly about worshipping the God of Christianity has become a multi-faceted object that strives to change people's minds on matters of importance' (2015, p. 278). The Mass retains some of the importance it has established as a vehicle for weighty reflection over the course of numerous centuries, but it is re-purposed in a way that reflects the ever-changing societal developments around it.

The spiritual dimension of the Mass has seen a similar process of development, and Rocke suggests that 'the masses to be discussed reveal that ideas about spirituality were becoming ever more complex, moving in an ecologically-driven, neo-pagan, post-colonial sweep, to embrace the entirety of creation, while not losing sight of monotheism and the meditational religions of the East' (2015, p. 242). Masses have increasingly become a vehicle for addressing a variety of different audiences – in particular as a result of their reinstitutionalization within a concert rather than a service setting. As Rocke describes:

> The secular concert venue nourishes the chalice of choice. It welcomes all who are prepared to abide by its rituals and its conceits, while expecting nothing more from them once they have paid for their ticket and entered the doors than their silence and their applause – at the appropriate times. The concert hall is a secular venue with a pluralistic attitude – a sounding board for ideas, not sermons; a place from which people can leave with new resolve or no resolve at all; a place for contemplation, for intellectual stimulation, or for simple enjoyment; a place for sacred music, a place for secular music, and a place for music that has no such impulse at all. Whereas the Church serves God, the concert hall serves music. Both, however, are communal spaces where humans gather for metaphysical sustenance. (2015, p. 277)

While both of the projects described in this chapter are connected in some way with the space of a church, in each case the space is either reimagined through its use as a concert or event venue or extended through the use of internet streaming alongside live performance. As I highlighted in Chapter 2, the role of 'secular' and 'civic' spaces in connection to Christianity, music, spirituality and religion has aroused increasing interest in recent research.[2] As a result of its open character, which is available for a variety of different publics and uses, secular space is fundamentally intersectional in nature, enabling the layering of and encounter between individuals' multiple and various identities, experiences and concerns, drawing together themes and realms of life in a way that is not always quite so easy within the realms of religious ritual (Porter, forthcoming). The format of the ecological requiem is one that has the potential to build upon the power of a similar intersection, drawing on the resources of spirituality, tradition and religious narratives while opening out on to a broader public and tapping into a realm of deeply felt existential concern.

Narrative and liturgy

Interviews with those involved in the different requiem projects demonstrated a creative engagement with the requiem format, both drawing upon existing formats and taking them in a variety of new directions. The format of a requiem for lost species, focusing explicitly on the consequences of the climate crisis in the creaturely realm, is something that has been explored in a number of different locations both as an act to raise awareness and as an act of lament. A commission for performance in Canterbury Cathedral draws directly on traditional requiem texts but supplements these with a range of others. Jonathan Arnold described the idea behind the work as follows:

> The original aim was that we could commission a classical piece of music for a traditional cathedral choir and organ ... it will involve a standard requiem text ... but those Latin movements will be punctuated with maybe a sort of Bach-style chorale or hymn for the congregation or audience ... there's also poetry ... So somehow all of these things have to fit together in short movements, some of which can be taken out and used as regular anthems for a church service ... this requiem is ... a bit more like a Britten *War Requiem*, in that it's about bringing people to account rather than just lamenting and saying we're guilty but actually, just as Britten

was saying, this should never happen again; we're saying this generation has to do something about it and actually change the direction of global warming. (Jonathan, interview with the author, 24 January 2020)

The commission is imagined as a resource and a symbol, as something that will raise the visibility of environmental issues within Christian liturgy or as a concert piece, and which can be used by regular parishes as much as by cathedrals in order to do so. The awareness of the connection between environment and different moments in the liturgical structure is an important one for Jonathan:

After the welcome and the gathering you go into a period of confession where we feel the guilt for our sins. And at the moment I think that's where the environment is sitting as an issue ... but we need to move beyond that now ... what we've all recognized is the importance of the requiem to lead the listener on a journey, which leads them from a sense of confession, perhaps a sense of, with the 'Dies Irae', there's going to be a day of judgement, a day of reckoning, but actually to end on a theme of aspiration and action and hope. (Jonathan, interview with the author, 24 January 2020)

The requiem, then, serves as a creative form that enables some of this integration with a longer liturgical journey, the space that a new musical commission opens up providing the opportunity to craft new narratives and lead participants along different pathways of experience from those that they might otherwise have embarked upon. Liturgy, in its movement between different moments of dramatic action, is particularly adept at showing the different entry points that issues of climate might have into Christian narratives.

Authentic emotion

As with the *Doxecology* project described in Chapter 1, the Canterbury Requiem has a certain chaos to its creative process. The initial artists involved in the project came and went as other priorities and concerns took over their schedules, and new ones entered with a range of different impulses, some of which made it through to the later stages of the creative process and some of which didn't.[3] Emma Pennington, who is Canon Missioner at Canterbury Cathedral, got involved in the project slightly late in the day after someone else who had been due to write texts for the project dropped out. Emma's involvement in the project arises not so much out of a burning desire to engage in environmental activism but, rather, as a desire to engage in a space of creativity and, at the same time, to process emotions and express things that she feels aren't often quite so readily offered a hearing in environmental activism more generally:

> To be really honest ... I wasn't writing because of the subjects. I was writing because I quite like writing ... The more I've got into it ... the more I've found it really ... inspiring to sort of have a voice ... Often, I feel that the voices around environmental issues are always very preachy ... and that just doesn't do anything for me ... What this has done is enabled me to actually ... just express those emotions which often I feel we're not allowed to express with environmentalism ... I suppose what's behind the reproaches and the sonnet of sadness is just a deep sadness, really, that the planet we love ... we're just so stuffing it up. So I think that's been nice for me to sort of enter into this issue, and find something that's a bit more authentic than preaching. (Emma, interview with the author, 11 February 2021)

While environmental engagement is not her primary motivation for becoming involved, Emma nevertheless carries with her a range of feelings around environmental issues, feelings that the compositional process has helped to draw out and express, thereby helping to remedy some of the alienation instilled by encounter with emotional regimes that don't always make quite so much sense for her. Emma's own attitudes and experiences go on to shape her desires for the work itself and for what it might have to offer people who hear it:

> I suppose what I would want to say at the end of this piece is folk to come out feeling that somehow doing something would free them to be who they are, truly are, and bring them closer to their relationship with God, and their relationship with his whole world ... to be honest, until you asked me, I never even thought about having any expectation of anything that people would get from this; it was simply two people having a go. (Emma, interview with the author, 11 February 2021)

Emma is wary of imposing any particular expectation or mode of feeling on other people; rather, she hopes that they will find some space for authentic engagement in the same way that she has also carved out some space to do this. Rather than hoping for them to internalize someone else's agenda as to how they should relate to the world around them, she wants them to find a place where they can be truly themselves and truly engaged both with nature and with God. Just like Jonathan, Emma emphasized the importance of the narrative dimension in interview:

> The best funeral services go through this kind of grief and sadness and memory and loss, and then it moves into a different kind of narrative and course and allows the person to let go, and then life to continue ... And so I think Adrian and I both wanted ... for people to leave not feeling that, yes, we're going to go out and ... sit outside our schools and not do anything 'til anyone listens to us, but ... that we felt empowered, that we felt life had been changed in some way in our approach. Not sort of just I'll give some money so that makes everything all right, or I'll go on a march ... but actually, in the way

they live their lives ... I think that people don't change their actions ... until they've started to change their understanding and their approach to things. (Emma, interview with the author, 11 February 2021)

This, then, is a space in which environmental issues are squarely on the broader societal agenda, and where creative spaces such as this requiem project arise as a result. A requiem like this offers both the creators and the audience an experience in the moment that is hopefully positive in some way, but doesn't necessarily have ambitions to reach out into particular kinds of practical action beyond the power of experience and encounter. Indeed, there is a sense that precisely this absence of a greater agenda might be what is needed for this project to be authentic. Goal-driven behaviour may help to achieve results, but it doesn't necessarily offer the human space for processing the changes happening around us that is needed along the way. In the course of the interview, Emma emphasized that she would not consider herself to be an environmental activist and is often quite disengaged from this kind of theme. However, the project seems to have offered a way into ecological engagement that makes sense for her, and a way of processing things. The space of the cathedral has a particular role in enabling this, as a result of its inclusive nature, offering space for a civic presence alongside Christian devotion, part of a national and state-connected religious landscape that needs to offer a possibility of inclusion and invitation to anyone who might turn up.

Imaginative possibilities

Emma's approach to the Requiem itself mixes a desire to draw on liturgical music with responses to specific aspects of the natural world and a creative engagement with a range of different scriptural and musical texts. It reflects a playful sense of the imagination where different influences can be juxtaposed, transformed or given voice, coming together into an assemblage

where tradition, imagination and events happening in the contemporary world encounter one another and come together to
create something new. Her writing takes the form of a kind of
experiment, trying out different ideas, mixing together different
sources of inspiration and drawing in different sentiments and
reactions that she herself has experienced, layering theology
together with different sentiments and attitudes to create new
spaces of meaning:

> I just threw on the page lots of different forms of writing
> ... sonnets, laments, reproaches, hymn, praise, all that kind
> of stuff. And thought about just the forms of that language
> would take ... I'm a big fan of Sanders's reproaches ... so I
> wrote the reproaches ... because I just liked the idea of God
> sort of going, you know, what have you done? What have you
> done to this wonderful world? ... And us just being sort of
> hurt, realizing how we've really messed it up in a big way ...
> My base sort of text was Genesis 2, the second creation story
> ... We wanted it to actually echo the requiem. And so the first
> part of the actual Requiem is going to have the Gregorian
> chant of the requiem ... 'Have you seen?' was actually 'Streets
> of London' ... looking at that sense of the sadness of some-
> one who's no longer connected with the world in that song.
> And then I was looking at this World Wildlife Fund website
> where it just had all these animals on the critical list ... most
> of which I didn't even know, you know vaquita – what the
> hell is that? ... it's like a little dolphin, and the dolphins are
> really having a hard time because of the fishing. And so many
> species are like just under threat ... And then the final one ...
> we're going to have 'All creatures of our God and king' ...
> something that updates that hymn to now ... The shrinking
> forests and the flooding, and the scorching of the sunshine,
> they are just being what they are. And in a sense, what they
> are, is to praise God ... even though it's been damaged, and
> it's broken ... we will put the original Victorian hymn into the
> minor key because that was written at a time of a great sort of
> imperialism and ... the beginning of the industrial revolution

... and then coming back to ... the major key ... trying to still give thanks for creation that we have at the moment. (Emma, interview with the author, 11 February 2021)

Adrian Bawtree's motivations seem slightly more pragmatic than Emma's: he is less cautious about pushing for practical action and happier to latch on to broader activist movements. He expresses a desire for reform that stretches from the individual, through the church and society to the political level, and is just as keen to rework the structures of musical participation as to see a politics adequate to the climate challenges of our time:

I took the inspiration a little bit from David Attenborough's last programme ... David Attenborough was telling us at the very end of that programme ... it's critical, but it's not too late if we act now. And I think that is the message I want to amplify, because at the end of the day, I know that if you just start telling people we're all doomed, they'll just switch off ... And I think that's why I was quite keen to get the two moves of the outer movements sort of right, while the opening is quite sort of dramatic and sort of Terminator landscape; the ending is, you know, green and pleasant lands and, I hope, uplifting. (Adrian, interviews with the author, 18 September 2020 and 25 January 2021)

Adrian's overall approach seems to be driven by a mixture of eclecticism and community, reaching out in a variety of different directions across the course of the project. Eclecticism is a way of drawing out meaning from different directions, while the community aspect is about bringing different groups together. On the community side, he wants the work to be something that people from around the church can sing together with school choirs, community choirs and other people from the life of a village. This ties into a broader vision for the direction that church music needs to be taking more generally:

It's about using music to bring groups that otherwise would not necessarily come across each other. And I think the church needs to do that. I think the church is supremely placed to do that. And it feels like the church has resisted it. I wonder how much the musical establishment also frowns on it … I think the inclusivity of music needs to be reclaimed. I think it's become quite exclusive … I think there are some really interesting opportunities for music here to emphasize social cohesion and challenge the notion that if you go down the community route you've somehow dumbed down the impact of music itself and that just isn't the case. (Adrian, interview with the author, 25 January 2021)

Adrian's imagination for the music is wide ranging, drawing on the darkness of some of Hans Zimmer's compositions, on hymn tunes and anthems, and on imagery relating to the four elements of wind, fire, air and earth. He talked to me about melting ice and its similarity to Poulenc's depictions of the guillotining of Carmelite nuns; imagery in Revelation; speeches by Greta Thunberg; Wagner's *Götterdämmerung*; the weaving present in Palestrina; David Fanshawe's use of recording together with live music in *African Sanctus*; biblical imagery of the angel with a flaming sword guarding the garden of Eden; David Attenborough; the narratives of scientists; a procession for an anniversary of Thomas Becket; and the musical *Jesus Christ Superstar*, among many other points of inspiration. There is an almost overwhelming variety of different possibilities stirring together that can evoke the imagination in different ways.

At the stage of our conversations, this swirling of different inspirations lacked a certain sense of coherence, but it demonstrates the rich variety of approaches that can be drawn on for a project such as this. Both Adrian and Emma collect together different narratives, different experiences and different possibilities, trying some of them out, abandoning others, and sometimes getting a little lost along the way. Their experimentation highlights the fruitfulness of an approach based upon narrative and imagery and the rich range of possibilities that can be evoked when using this as a basic model.[4] What is striking, however, is the rich imaginative journey that both individuals embark upon. Both use the requiem as a chance to think through issues of grief, hope, humanity, creation and destruction and, in seeking ways through this, the format of the requiem allows them to reach out imaginatively into a vast range of different areas of inspiration. Analogies in other works and other realms help to get an emotional, narrative and imaginative handle on the issues with which they are being confronted. There is hope that this imaginative journey might be one that others can also embark upon through their composition, and they look for points of entry through which others might be able to join them on this journey.

Justice and humanity

Song of the Prophets takes a different approach from the Canterbury project, with a closer focus on human communities, and a more tightly controlled four-movement structure. The project is the result of a commission on the part of Christian Aid in collaboration with the Chineke! Foundation and St Paul's Cathedral. The involvement of Chineke! as an orchestra formed of black and ethnically diverse musicians, as well as the range of musicians and composers invited to collaborate, helps to make visible the crucial importance of non-white experiences and perspectives both in Christian Aid's broader development work and in relation to climate change in particular. The Christian Aid Requiem has a clear activist element to it, commissioned by a charity working in areas of the world that are already suffering as a result of climate change. In the introduction to the programme, Christian Aid Chief Executive Amanda Khozi Muwashi writes that 'Tonight, you are going to be part of something special. An evening that uses beautiful music, gifted musicians and passionate people to draw all of our attention towards the biggest crisis facing our world today: the climate crisis … Listen well, feel deeply and afterwards join us as we come together to tackle one of the greatest injustices that we face' (Christian Aid, 2021a, p. 6).

The musical compositions themselves are somewhat brief in nature, specifically limited in the commission to two minutes each. However, the event itself, which ultimately took the form of an internet livestream, lasted almost three-quarters of an hour, interspersing more spontaneous performances with speeches from a range of different voices, including the former Archbishop of Canterbury, Rowan Williams, alongside climate activists from a range of different countries and the leaders of Christian Aid and Chineke! themselves. Structurally, the Requiem sets aside any reference to the traditional movements of a requiem. The four movements, Creation, Ruin, Recovery and Redemption, were each handed to a different composer, and each is described in the programme with a fuller brief of what the section hopes to achieve:

Creation: This is the opening section of the piece and takes its inspiration from the creation poem in Genesis 1 in the Bible. 'In the beginning when God created the heavens and the earth, the earth was a formless void and darkness covered the face of the deep, while a wind from God swept over the face of the waters. Then God said, "Let there be light"; and there was light. And God saw that the light was good; and God separated the light from the darkness. God called the light Day, and the darkness Night. And there was evening and there was morning, the first day.'
Words: Growth, emergence, birth

Ruin: The once perfect world we are introduced to in the opening 'Creation' section is now ruined. The world is on fire. Climate change is having a devastating impact on every area of life – from basic needs such as food and shelter, to issues such as education and women's rights. Some of the hardest-hit communities are having to change and adapt in order to survive. They are facing the full force of the droughts, winds and storms that are increasingly common because of the climate emergency.
Words: catastrophe, destruction, disaster, death

Recovery: Around the world, Christian Aid is helping to restore the dignity and livelihoods of those in communities that have suffered recent storms, droughts and floods. An estimated three million people were affected by Cyclone Idai in Mozambique, Malawi and Zimbabwe in 2019. The cyclone saw at least 750 people killed and an estimated 400,000 displaced. In the immediate aftermath of the cyclone, Christian Aid was there to help those in remote areas who were desperately in need of food, water, clothing, shelter and medicine. Today, we are still there, rebuilding homes and livelihoods long after the humanitarian crisis has disappeared from the news. People who care about the dignity, equality and justice of all should join together to help our global neighbours recover in the aftermath of ruin.
Words: aid, elevate, lift, clear

Redemption: This is the finale of the piece and points to a hopeful future in which all things will be made new. It acts as a rallying call; a future to which all of humanity can strive for the sake of all. It should be joyful and hopeful.
Words: renewal, rebirth, restoration, hope. (Christian Aid, 2021a, p. 9)

The performance is undergirded by a theological publication produced by Christian Aid in May 2020, which reflects upon the inequalities and injustices of climate change, using the lens of prophecy to offer both critique and hope as well as a challenge to the wider society. According to the report, prophets help us to reimagine the world, rooting their hope in a God whose promises cannot fail, and the report urges us to hear the insights emerging from the global south, to move to action and to listen to the sense of God's loving purposes in Scripture (Christian Aid, 2020, p. 4).

Is redemption really possible?

Interviews with two of the composers involved in the project offered very different descriptions of what it meant to be involved, and what it means to approach this kind of commission. Both came from different backgrounds in relation to composition and to questions of faith and activism. Both were clearly interested in coming on board with the commission process, but both also had different expectations of and hopes for the potential outcomes of their involvement in the project. Roderick Williams, who composed the final movement of the Requiem, suggested that he came to the composition as someone who was 'concerned, interested' and who does what he can but who had 'never been strongly politically active' (Roderick, interview with the author, 10 February 2021) when it comes to questions of ecological concern. He is impressed by the activism of others, and participates when there is the right opportunity, but doesn't typically extend his activities beyond that realm. Likewise, he comes to the project without any strong religious motivation, and much more of an interest in the musical ideas he can draw on and play with over the course of the writing. As such, his involvement demonstrates something of the broad coalition that both work centred on the climate and classical music performance are able to bring together, enabling different individuals to find the points of access and relationships to projects with which they feel most comfortable as they reach out in different directions and become part of individuals' lives in a variety of different ways. Roderick explained his approach to the composition to me as follows:

> They'd given the four composers ... a very specific brief, I think mine was redemption, and ... they asked, could it be in C major, please? ... if you're going to be redeemed C major might be a good key to be redeemed in ... What I suggested was that I tie the piece up as much as possible by using the material for my three fellow composers as a sort of grand finale rounding off ... In fact, when you hear it ... you will hear

immediately that what caught my imagination was the request that it be in C major. So I simply took the piece of music that immediately popped into my mind as being synonymous with C major, which for better or worse happened to be the opening of the *Meistersinger* overture ... I just hear the big C major chord as a scale that goes off, the highest instruments go upwards on a rising scale while the bass instruments descend on a scale. And so I simply took that shape ... and went off on one with it ... It's actually only talking to you about it now, that the extra connotations of that particular work and where it stands in western classical art, an irony of it becomes more delightful to me ... that piece as a particular work has resonances that I hope people will bring with them when they listen to my expansion of it. And those references are useful to me, actually ... Now, for me the idea of redemption. In our world today, that kind of leaves me with an artistic choice ... Do I leave the audience feeling they have been redeemed? Or do I leave it open ended in a question mark? Because, just the type of redemption is ... open to interpretation, 'Are we worthy as a species of redemption?' ... it's like ... the end of the Mozart *Marriage of Figaro*, where everything is resolved and a happy ending. We're all left with the feeling, well, yes, he got to the last bar in the major key, but I'm not quite sure ... I've ended on a very big, loud C major chord. It will be up to the audience to fathom whether that is in some way trite ... I do feel that it's quite shiny, and quite glitzy. I'm not quite sure whether it's necessarily profound. And that, I think, is my final sort of statement in this piece, that we can put a large, shiny piece of sticking plaster over the whole thing, but whether it's actually solved is another question. (Roderick, interview with the author, 10 February 2021)

Over the course of the conversation, initial technical and artistic parameters that served to give the compositional process impetus and form give way to a more profound question of the nature and possibility of redemption. As a composer, Roderick is wary of offering a final answer to this, but instead opts to leave the audience with the question of whether the efforts of humanity are really enough. In interviews regarding music and podcast projects focusing on themes of climate change, the musician David Blower raised similar questions of what it might mean to create a narrative that doesn't necessarily have an easy point of hope at the end of it, but instead sits with the tension that our existing narratives might not be as secure as they once were and we might therefore need to find a different emotional space that encompasses some of that trauma:

> I don't think that human beings have contemplated human extinction. It hasn't really been something we've thought about. But then when you've got like the Cold War, and then climate change, you start to think, 'Oh, well, could we? Could we make the world uninhabitable for ourselves?' And this is quite a strange thought to religious narratives. Because I think most Judaism, Christianity, Islam, most of these kinds of faiths assume a final curtain when redemption happens. And when that final curtain pulls, we don't imagine it being pulled across a humanless planet, so there's a bit of a trauma to our theological imagination in thinking about, 'Oh, well, what if we, what if that happens?' I remember somebody saying it was a trauma to the Christian imagination when we realized that the Earth had been around for millions of years before humans were on it. And now we've got the other bookend of that trauma, where we're thinking what if the planet's around for millions of years after we've all died? Does it create theological problems? Well, not necessarily. But it's just been outside of our imaginary and when you bring that on to the table, it's destabilizing. But if you don't, then you sort of live with this niggling feeling that there's something important that you're ignoring. And it bothers you from behind the door, doesn't it? (David, interview with the author, 26 October 2020)

While these kinds of sentiments have a certain tone of profundity to them, Roderick is keen to avoid the burden of being too earnest in his endeavours, perhaps sceptical of the very possibility of such earnestness in the face of bleak realities with which humanity is currently left to grapple, but also in the face of a collaboration with other composers who have dedicated a greater proportion of their careers to the craft of composition:

> It starts with this rather bold and brash quotation from a Wagner score from a piece that is the quintessential German West, landmark of all that is Western and WASP [White Anglo-Saxon Protestant] civilization. I have a feeling it might cause the odd smirk. But that again is deliberate ... I have a feeling because, again, I'm a practising musician, and I spend my time among practical musicians so that possibly nudges me towards the ironic and the sceptical, and the practical ... a lot of practicality comes into the music I write, even if it's only the knowledge that Chineke! will not have much rehearsal time for this piece. (Roderick, interview with the author, 10 February 2021)

While Roderick contextualizes his sense of humour in his broader life and attitude as a musician, foregrounding its connection to his particular life story and circumstances, this is something that he strategically adopts and utilizes in order to play with the audience and their expectations:

> For me, it's the managing of expectations, that when an audience sits down to a piece of music written by four contemporary British composers, they will have an expectation of what music they want, that they might be about to hear. And I suspect that with the other three composers, those expectations may be met. With mine, when they hear the opening of Wagner *Meistersinger*, I might be able to wrongfoot them briefly, or maybe they know that I have this, I don't know ... you could call it a warped sense of humour. (Roderick, interview with the author, 10 February 2021)

The humour that Roderick describes has a strong personal dimension to it, but it is not without broader strategic significance. Mark Pedelty describes a dual role for humour in environmental activism, suggesting that 'Humor and narrative are both self-protective and effective, much more likely to hit home with skeptical audiences than finger-pointing or deadly seriousness prose' (2016, p. 55). While Roderick emphasizes this protective dimension, there is also a sense that he doubts the effectiveness of serious musical strategies and believes that humour can have the power to jolt the audience into a different way of thinking. Personality and situation come together in order to offer something that has the chance to create some sort of effect in the room, even if that be a simple momentary glimpse of surprise.

Activism and solidarity

In contrast to Roderick, Shirley Thompson came to the commission both with a long-standing commitment to musical activism and with an approach to such activism rooted in a personal Methodist faith:

> A good portion of my music is about highlighting issues within social affairs, current affairs, political affairs, historical issues … I see music, the arts, as a way of bringing another perspective on issues that need to be raised and to be examined. So for me using music as a campaigning tool … and I see it as a way that I can be of service. I think, for me, as a Methodist … actually, social justice is why I was drawn to Methodism … I think [the Requiem] will get a lot of attention in the spaces that it needs to and will provide a legacy, and that's what you can do, especially classical music … for people to see this point in history, what happened, and how effective it was that we've come together to do something like this … When I first started, about 20 years ago, I was accused of writing music and politics and people were saying, 'Oh, gosh, you're going

to be blacklisted for doing the things you do' ... And now everybody's doing it ... maybe because I've got a film-making background, I've always thought, how can we use music to tell a story and to highlight issues. (Shirley, interview with the author, 26 April 2021)

This broader activist involvement sets her up to have different expectations of the composition from those of Roderick. Shirley's activities have given her confidence that the work can create a broader impact, it can draw attention to an issue and it can stand as part of a longer-term historical trajectory, a sense of trajectory that seems also to spill across into theological narrative and the different stages of humanity's relation to God and to the world. Her own particular contribution is rooted in very specific recent events and is strongly anchored in a certain imagination of present-day reality:

Mine's about recovery, so the way that we can help aid, elevate, lift and clean, and clear up after landslides and the devastation caused by floods and drought ... And just looking at how just thinking through how countries have recovered after major catastrophes ... And I've tried to, in my piece, reflect the devastation of what happens ... physically and emotionally ... and then, through the work, end with something that's reflective, but reflects on the stoicism of the people that are recovering and trying to rebuild their lives ... For the last four or five years, I've been working with Kenyans in Nairobi. So I asked them about the major landslide that they've had ... I asked my friends that I have there what the effect has been. (Shirley, interview with the author, 26 April 2021)

Her piece is driven by her own involvement and connections with particular people and particular situations, alongside a strong sense of transnational social solidarity. It focuses on the human effects of climate change and pays attention to specific perspectives and experiences of others, not assuming too strongly in advance what their experiences might be supposed to look like but drawing on the particular. This particularity opens up on to a broader global imagination:

> I think it's all about interrelations ... we are humanity. As much as we build up all these walls. But we're all interrelated. You've got the monsoons in India but become hurricanes on the other side of the world. And that's physically how we are so interconnected. The sun rises on one side of the world, but it sets on the other side of the world. So, for me, it's all about the connectivity of us, in our politically divided spaces ... these borders that have been drawn up are all very recent, certainly in Africa. (Shirley, interview with the author, 26 April 2021)

The Earth and our relatedness are prior to human politics, and a problem in one place almost inevitably connects to and spills over into situations happening in other locations. This connectedness underlies both the need for solidarity and the possibility of action. Shirley is highly pragmatic and strategic in her use of music, a pragmatism that is also undergirded by a stronger confidence of divine involvement in the world:

> I'm very much about immediacy and reaching a public. So big public events, like this, what excites me is that you're pulling in the general public to an event like this, and it is going to affect a lot of people ... I want people to be moved by my music. I want people to think about what I've written, maybe why I've written it ... I hope with music ... to take people to the place with the music so they can feel some empathy with what's going on ... You know, this is happening, it's very far from you, you might not be experiencing it yourself. But here it is, you might have some kind of affinity sometimes in relation to it. You might find yourself in a similar situation ... I have a strong belief in God's work ... So whatever that work is, then I try to follow that lead ... I'm just a channel, you know, I'm a vessel. (Shirley, interview with the author, 26 April 2021)

Music can generate emotional solidarity, working to build up empathetic relationships and helping to create a sense of shared experience across distances that might otherwise be challenging. While Roderick strategizes his relationship with the audience through humour, Shirley does it with empathy. Both are keenly aware of the possibility of creating a reaction, but their entry at different points in the requiem narrative as well as their differing approaches to musical composition more generally lead to differing imaginations of what their musical composition might be capable of. In keeping with requiem Masses in general, neither dwells particularly long in a space of grief, but both use a moment of loss and destruction to open up on to broader narratives and possibilities.

Where are we heading?

In seeking to process sorrow and loss, the question of trajectory and ending is a crucial one. There is a deep awareness of the inadequate emotional journeys that it is possible to lead an audience along and a sense that, at this stage, there is mediation to be done between hope and despair, between recognition and protest and between different possible futures that still lie ahead of us. To a certain extent, this trajectory must currently remain an open one. We do not know how far the climate crisis will go, we don't know how many species we will allow ourselves to destroy before we find a way to turn things around. This does not prevent composers from pointing to the different possibilities ahead of us, however. The very need to end a requiem somewhere forces us to contemplate the narrative continuations that are available to us, showing us our own role in this story of grief and destruction and inviting us to take on new relationships with this story as we head forward into the future.

In conversation with the various composers, we can begin to see some of the different possibilities that the requiem format opens up. Climate requiems are complex phenomena. They represent grief but also hope, capitalizing on established forms but taking them in a range of different directions, hoping for action but never assuming they can bring it about on their own. As a liturgical form, they often take little from the requiem Mass itself, but they nevertheless take people on a journey that is only made possible through the use of multiple sections and movements each with their own characteristics. Their connection to the art music canon enables ecological requiems to evoke and draw upon a rich range of further narratives and imagery, and they are both expressive of the climate in which they are forged and an attempt at intervention within it. They are a space in which a variety of different ideas, directions and influences can be brought together, resulting in a similar openness to the different possible effects they might have on members of a listening public. At the same time, they are positioned on the boundaries of institutions, at the intersection of public, church and musical bodies. Part of the role of those creating them is a process of mediation between competing demands and points of critique and inspiration. Who needs to hear what from whom? What new alliances can be created? Which possibilities remain just a dream, and which ones can truly become a reality? The lack of concrete answers does not take away from the power that these questions pose but, rather, asks us to grapple further in a space where we are far from having all the answers.

Notes

1 For example, the *Requiem for a Dead Planet* held in London on 19 July 2019.

2 See, for example, Arnold, 2016; Brown, 2007; Heaney, 2013; Ingalls, 2012; and Klomp, 2021.

3 Indeed, at the time of writing it remains somewhat unclear to me whether the work will make it through to any kind of final complete performance. An initial movement was released online, with the promise of more; however, these have yet to fully materialize. In this sense it is a project where the creative process and ideas involved are more central to my narrative here than the analysis of a final performance.

4 Partly, I suspect, it is this richness of possibility that has led to delays in bringing the work to fruition. When there are so many possible options, settling on one and moving it to completion can be a somewhat overwhelming endeavour. However, this is far from the only reason for delays in the composition. Challenges of funding, managing multiple projects and jobs, gaining the necessary institutional support, and juggling composition as a secondary task alongside primary employment elsewhere all contribute to the path that the composition has taken.

5

Sound and Natural Environments

Music and sound were not an easy and uncontested part of my first visit to a Forest Church. Meeting in a public car park before a short walk into the woods, it was clear that there were different opinions as to whether or how we might want to sing. Does it make a gathering too churchy? Does it alienate? Does it establish a sense of sacredness and boundary? Do we want boundaries in a group that is open to a range of different participants, and which opens out on to the natural world? We began with a sense of uncertainty as to what was appropriate or whether singing was really a good idea. The gathering itself begins with a short chant, sung by the vicar, but music is absent from most of what we do. We hear thoughts from different individuals before being sent off into the forest on a search for things we might want to bring back. As I search around for pine cones in different conditions – fresh, half-eaten by squirrels, decayed, broken – I am conscious of sound and speech largely through the feeling that it is better not to speak or converse when trying to become aware of the forest around me. Towards the end of the gathering, however, music once again enters the frame through a small electronic device that we are told will convert the electrical signals of plants into music and sound, helping us find a way of perceiving their life and activity and perhaps to connect with them in a new way (Music of the Plants, n.d.). There is both scepticism and curiosity in the broader group. Many members are curious to attach the electrodes to different kinds of tree, wanting to experiment and find out if they get different results. Some are also curious to find out how it works, slightly unconvinced that it really translates the signals very directly, and wondering what exactly it might be translating. The vicar asks afterwards whether any of the group feel more connected to the creator, but there is some doubt as to the real authenticity of the connection on offer, and conversation seems to turn more to practical questions than to spiritual experience.

Tuning in to nature

While the previous chapter focused on a musical form deeply rooted in historical Christian traditions, not all groups are quite so convinced that Christian ritual practices form the most helpful point of departure and, in line with wider suggestions that indigenous epistemologies might be an important source of wisdom in grappling with the challenges of environmental crisis (Titon, 2019, p. 103), Forest Churches lean in new directions precisely out of an awareness that something important might be picked up from elsewhere. In comparison to the other groups with which I made contact, Forest Church is the least directly motivated by climate change. Rather, it is part of a broader impulse to reconnect in some way with the rhythms of the natural world. According to the Mystic Christ website, 'Forest Church is a fresh expression of church drawing on much older traditions when sacred places and practices were outside – but it is also drawing on contemporary research that highlights the benefits of spending time with nature in wild places' (Communities of the Mystic Christ, n.d.). Forest Churches follow diverse patterns and have diverse motivations, but they are centred around a common desire to meet in outdoor spaces, and in some way to explore the spiritual possibilities that these places offer, sometimes in close relationship with an existing congregation and sometimes as a freer initiative that attracts anyone interested in this exploration. Some meet regularly, some only a handful of times a year, but they find common ground in their exploration of outdoor spirituality and in their openness to the different ways that outdoor environments are able to shape that spirituality in ways that are more than simply taking indoor forms of worship and reproducing them in a different location.

Forest Churches are, of course, not alone in seeking in some way to reconnect with nature and, just as the website suggests, we see a range of understandings of this experience offered up in recent literature, some of which focus on well-being and others on spirituality. Bron Taylor (2001a, b), for example, offers a helpful overview of some of the different nature-based spiritualities that emerged up to the turn of the millennium, both on the level of individual experience and of wider activist or more explicitly spiritual movements. He describes the emergence of deep ecology and radical environmental movements in the second half of the twentieth century, and a variety of New Age movements and earth-based spiritualities that followed on from them, alongside more recent movements connected to bio-regionalist thinking, wilderness rituals and magic. We see these spiritualities both gaining in popularity and taking a wide range of different forms over the course of different decades. Peter Ashley discusses the attraction of wilderness spirituality on a more general level, suggesting that it involves 'a feeling of connection and interrelationship with other people and nature; a heightened sense of awareness and elevated consciousness beyond the everyday and corporeal world; cognitive and affective dimensions of human understandings embracing peace, tranquillity, harmony, happiness, awe, wonder, and humbleness; and the possible presence of religious meaning and explanation' (2007, p. 65). Lia Naor and Ofra Mayseless (2020) have likewise highlighted the therapeutic qualities of spirituality in nature, drawing attention to its physical and sensual embodiment of spirituality, a connection between the immensity of nature and experiences of a higher power, the experience of interconnectedness and belonging that can arise in nature-based spiritual experiences, and the ability of experiences in nature to reflect, mirror or echo personal information, leading to processes of self-discovery. Other authors make similar suggestions. Paul Heintzman (2003), for example, focuses on the socially enabled and therapeutic characteristics of spiritual experiences in nature, while Tristan

Snell and Janette Simmonds note the way people are often drawn to protect environments in which spiritual experiences have occurred (2012).

A rediscovery of nature-based spirituality sits alongside a growing interest in the value of nature connection in other areas of life and research. A group of authors writing for the *Annual Review of Environment and Resources* (Russell et al., 2013) focus on the connection between nature experiences and well-being. The authors draw out a range of different dimensions, looking in turn at ten different categories, including physical health; mental health; certainty and sense of control and security; learning and capability; inspiration and fulfilment of imagination; sense of place; identity and autonomy; connectedness and belonging; and subjective well-being, alongside a small subsection devoted to spirituality. Capaldi et al. (2015) push in a similar direction, while other authors (Barbaro and Pickett, 2016; Gordon, Shonin and Richardson, 2018; Schutte and Malouff, 2018) focus explicitly on practices of mindfulness as a mode of encounter with nature. It is clear that encounters with nature can be understood in a variety of different ways, and that potential connections with spirituality sit alongside a range of other experiences that people seek or undergo in these outdoor environments. Outdoor environments are often tied together with a search for connection or well-being, but this can take a range of different forms, and it is hard to pin down into a single narrow philosophy of how this should happen or what form it should take. Rather, these spaces are open to different ways of thinking or experience precisely because they exist at a certain distance from the kind of frameworks and institutions that push us to understand a space in more specific and pre-defined categories. So how do music and sound emerge in Forest Church spaces? How are they related to the broader range of nature-based spiritualities, activities and journeys? And how do they participate in a wider reimagination of human–nature relationships?

Rituals, rhythms and relationships

The broader Forest Church movement is a diverse one, and music is far from ubiquitous. At Ancient Arden Forest Church, however, music is a regular part of gatherings. *The Song of the Wheel*, a songbook created by community member Alison Eve (2015), focuses largely on a musical tuning in to the natural rhythms of the year as articulated within pagan calendars.[1] Alison explained some of the thinking behind the songs:

> I had in front of me from my resources a number of pagan rituals that I was looking at, and looking at the flow of the ritual, the elements that were included, and then thought about the ways in which they could be re-crafted and reimagined ... so in Wicca it's very much about a sort of a duo-theistic, you know the lord and the lady, the god and the goddess ... I re-crafted it to calling the divine feminine and then calling Christ, so reimagining the wild lord as Christ.
>
> The first bit of singing that we usually do is weave the circle, and it's very useful there because you know the whole idea of weaving the circle is that when you go outside you haven't got church walls, so you're creating a temporary temple. It's a very porous one, it's very open, but it's just a space that you create and the songs that we use at that point are very powerful at helping weave that circle along with the use of elements such as water and incense and things like that and walking around the circle ... after that ... we usually have a song for the lighting of the fire ... fire is creating the warmth of our circle of fellowship ... and obviously we included in that fellowship not just the humans that are there but all the people who share that space, even if they're not human people.
>
> It's marking the seasons in a much more direct way than we do in church ... and very much noting the environment in which you're in and connecting with the space ... you need to go there and ask permission to use the space ... the import-ance of paganism in the whole climate change thinking is it's

coming from a different perspective, it's coming from the perspective of asking the trees for permission, of recognizing our
participation in creation rather than being in this place above
the rest of creation ... taking time to listen to the smallness of
the space – you know, to listen to a leaf. (Alison, interview
with the author, 21 January 2020)

It is notable, here, that song and music are far from self-
contained; rather, music becomes a means of relating to the
rhythm of the year, an accompaniment to ritual action and a
means of establishing relationship. The issue of power and hierarchy is also important and is something that continually has
to be renegotiated in understanding Christian relationships to
creation. Here, dynamics of dominance and exploitation can sit
alongside those of stewardship and care or of partnership and
integration. Music and sound is an important medium through
which these relationships of power can be renegotiated. If music
is thought of as a medium through which relationships and
interactions are established (Porter, 2017; Titon, 2019, p. 112),
then the move towards musical activity in the outdoor world
raises important questions as to the significance of this transition for the kind of ecological relationships that are formed in
and alongside music and song. The movement outdoors carries
with it a reworking of musical practices, and these contribute
very directly to the changed power relationships between those
singing, the world around them, and one another (see also
Ryan, 2019). According to Simon, another musician within the
community:

Musically and instrumentally it's a case of stripping down,
which then I think puts an emphasis on ... the vocal line ... I
think for me the focus really falls on what the voice is as well,
when you're outside ... a part of what I think Forest Church
is doing against this background of a reawakening to nature
is about saying, well, not everything is about being inside ...
it's not the only way to do it, and a lot of the things that we've
built up around us to create this sort of indoor lifestyle are

the very things that we're saying we've got to start undoing in order to tackle things like climate change ... the singing that we do enables you to concentrate and focus a bit more on the tone of what you're producing, you know, the level of volume at which you're singing, you know, is that appropriate? (Simon, interview with the author, 10 February 2020)

This attention to level and to voice is a crucial one since it is precisely this kind of relationship that often configures relationships of power and energy between actors and spaces in the act of singing (Warren, 2021). Music becomes a means through which the balance between humans, other humans and the environment is negotiated; and as the climate crisis suggests the need for a rebalancing of this relationship, so in a capella singing in an outdoor environment this rebalancing also takes place in song. Level isn't the only way in which this balance of voices between humans and other actors can be imagined, and Barbara, for example, drew upon the idea of conversation as a way of imagining the different back-and-forth interactions that take place:

I remember a meeting where we were having a discussion ... and then there was a sound of wind in the trees, and Bruce said the trees are joining in with our conversation. And that's not a musical event as such, but I think what we're doing is – for some of us anyway – music is such a natural way of responding. And it's a shared way of responding, isn't it? (Barbara, interview with the author, 8 October 2020)

She highlights the way in which the outdoor space gives a new dynamic to the dialogue between different actors, a dialogue that is bound up with a fundamental sense of ambiguity between divine and natural worlds:

Barbara: When you're singing in an indoor space, obviously, in a worship context, we remind ourselves that we're singing for God ... But actually, realistically, what we're doing is listening to one another, and God is part of the equation, but we're singing for one another. I think when you're outside, it's much more obvious that you're singing for something else and not for the other people who are there ...

Mark: Can you pinpoint more closely that something else?

Barbara: Um, no, I think that's the thing ... it's precisely how we're not choosing to define our sense of the divine that makes Forest Church an open place where people on different faith paths can meet; it's precisely because we don't say, we're singing to God, or we're singing to Jesus, or we're singing about Jesus, that people feel able to meet in that space.

Mark: But it's a sense of the divine more than a sense of the natural world?

Barbara: It's both, so Forest Church is a place for people who sense the divine through the natural world ... It's not just about nature, but it's certainly not just about the divine. It's the two together. (Barbara, interview with the author, 8 October 2020)

There is interaction here between a whole range of different actors, and this interaction is bound up with a sense of openness that almost inevitably ends up extending beyond those explicitly addressed or engaged. The interaction reaches outwards without being bounded by any particular limit and, in this openness, there is the potential for the divine to emerge precisely because the sound is not addressed in one particular direction.

Adapting traditions

While Alison adopts pagan traditions as part of her musical practice, Barbara focuses more strongly on the adaptation of pre-existing hymns and folk songs. The use of familiar repertoire is partly pragmatic in nature, designed to draw on things that people are going to be able to join in with easily, but also involves a sense of connection with local tradition and her own ability to adapt lyrics and liturgy more readily than compose completely new songs. This intersection of folk tradition and Christianity can often involve singing about the saints – a topic that, as a non-conformist, feels somewhat transgressive for Barbara. But Barbara finds that it can be a means to a more inclusive, less patriarchal Christianity. In a similar manner to Alison's focus on the rhythm of the year, Barbara's reworkings often focus on the relationship of songs to particular seasons, places or occasions. A reworking of 'Now the green blade riseth', for example, maintains much of the imagery of the original but adjusts it for a service focused on the theme of Viriditas – after Hildegard of Bingen – emphasizing the green power of God and the greening and verdancy of the Earth. Barbara suggests that this was a matter of:

> bringing it further forward into the year because actually that greenness starts appearing at the beginning of February, that's when we see it. And so one of the things I tried to do in my writing is to focus on the actual plants and animals that are around us in our space so that we're singing, I'm singing about lichen and I'm singing about snowdrops and birds and things that are actually in the space that we are that we're singing in. (Barbara, interview with the author, 8 October 2020)

An additional verse for 'Jesus Christ the apple tree' was likewise introduced in order to give it an autumnal rather than Christmassy seasonal focus. This is born out of an awareness that, while the Christian tradition has seasonal hymns appropriate for various times of the year, there are many that it currently neglects:

> Our Christian hymnody is partly seasonal, and we have Harvest hymns that we sing at Harvest and Advent hymns and that kind of thing. But if you're trying to recover the focus, the eight-point Wheel of the Year, then there aren't specifically hymns that focus around those points – or not hymns that are

accessible to a modern audience. (Barbara, interview with the author, 8 October 2020)

The stories, connections and associations of particular hymns are important to Barbara, and a hymn that she reworked for Michaelmas gives a flavour of some of the different considerations that can come together in one individual song:

> We were doing a Michaelmas-focused Forest Church. And I kind of knew this hymn … But when you look at the original origin of it, it was written by, or it was overheard by, a guy who was going to visit a bunch of slaves who had been emancipated. But they'd been emancipated during the Civil War and they were on an island, and … the only way that they could get off it was by rowing … and … the original lyrics are absolutely fascinating, but they didn't rhyme. So one of the things I did with it was that I made it metrical and gave it a rhyme scheme all the way through. And we were doing it … in the churchyard of a local village, where the guy who had built the church … had made all his money from the slave trade. So I felt that that was an appropriate response to be doing it in his churchyard. (Barbara, interview with the author, 8 October 2020)

A sense of tradition, place, time, justice, individual actors and social history come together to produce a highly contextual form of meaning and significance. If Forest Church thrives on attentiveness to particular spaces, then this seems to take such considerations somewhat further, relating them to a longer-term human presence in the landscape, one with which a connection can perhaps be maintained through the ongoing continuity of traditions, places and physical structures.[2]

Freedom, play and welcome

There are diverse ways of doing church outdoors and while some Forest Church groups are driven primarily by a connection with nature or folk or pagan traditions, sometimes questions of demographics and who might come along come further to the foreground. Pete's desire to set up a Wilderness Church connected to his parish in the USA draws on communal experiences of youth groups growing up, and hooks into the sense of the outdoors as a place where there's more freedom to engage in different practices, and where there might be lower barriers of entry for individuals who might not normally set foot inside a traditional church building. This orientation is closely tied up with experiences of a broader range of music, which in some sense manages to overcome too strong a divide between the sacred and secular worlds:

> The foundation of my faith came from the summer camp that I went to ... singing outdoors is something I grew up doing ... Peter, Paul and Mary, Bob Dylan, the sort of 'Peace train', a lot of Cat Stevens, as well – as you know – there's a spirituality in what we would call secular music that I don't see as much in today's music ... I lived in New Orleans for a while ... and my church was a very high Anglo-Catholic church ... It was very diverse, economically, racially, and it was about 50 per cent gay, but we sang. So you had this very high liturgical worship, but we also sang gospel songs ... And I really loved that, that we sang the songs that you don't really sing. They're sort of in the Christian canon, but in the Episcopal Church, at least, we don't sing them ... I think that while there is some evidence that younger people especially are looking for more of a high liturgy, I think that the barrier for entry is also pretty high ... And part of what I am hoping to do with Wilderness Church is to lower some of those barriers for entry ... without losing the sacredness of what God has given us in our tradition. (Pete, interview with the author, 20 December 2021)

In starting a Wilderness Church, Pete wants to sing the kind of songs that aren't possible within the bounds of his regular community, and he seems to be aware that this broadening of musical repertoire is also connected to a broadening in the kinds of people who are likely to come along. Similar themes are common to a number of Forest Church groups, who seek to establish Forest Church as a place for a range of different people to come along to regardless of their own particular commitments to Christianity or other faiths. Music and sound play a key role in shaping the nature of that welcome and how different people feel comfortable participating in an activity with different levels and types of expected participation and different kinds of ritual boundary – or lack of ritual boundary – around the meeting. Outdoor worship can reach out in other directions too, and recent years have seen an increasing proliferation of outdoor groups aimed at children or families, such as Mossy Church or Muddy Church. Sam's description of outdoor worship highlights the way that an openness to different demographics can be an integral part of the outdoor dynamic:

Mark: So what is the vibe of outdoor worship, then?
Sam: I think it's much more in the round rather than front led … I think it's instinctively all-age, rather than kind of aimed at one particular age group. I think it's trying to be shaped by the seasons and the natural world around us and be kind of indigenous … There's more around both ecology and science in general. There's more about thinking about the planet and natural processes … and we're quite conscious of wanting to have Bible content, but it tends to be Bible content that is also quite strongly related to natural processes and things. (Sam, interview with the author, 10 January 2022)

This can cross over into dynamics of play, and a broader understanding that spirituality, play and other activities don't necessarily need to be split into separate categories, but often work fluidly together:

Our theme yesterday was play, we were thinking about, sort of jumped a little bit off epiphany, Jesus being a child, Jesus growing naturally, like children naturally grow. And then so that led us on to spend a session on play ... We've always incorporated play and have seen, you know, it's not just about engaging kids or keeping them involved. But it's been about blurring the boundaries between 'all of this is a sacred thing', and then 'this is a secular thing where you play' ... and ... I think there's always that kind of idea that this play is inherently a good vessel for spiritual growth and engagement with God. (Sam, interview with the author, 10 January 2022)

I think that spirituality and play are almost so close, I can't tell them apart ... and if you're talking about how children learn, and how children learn through playing, there's a state of play that children get to called flow, which is where they're really deeply into whatever it is that they are exploring at that moment ... that you can't disturb them ... they're just in that zone. And I think that that kind of flow state feels very much like prayer to me, that kind of deeply being inside something else, being kind of outside of myself almost. And that's where exploration and growth happens. (Rachel, interview with the author, 5 January 2022)

Flow, here, serves as a kind of mediating category between prayer and play. The kind of states that are attainable in both activities are similar, involving a certain kind of attentive, absorbed engagement with an activity. Both are about a certain attitude or relationship with the other, involving a transformation of the self and a wider process of growth and change. This interrelation means that both activities can be cultivated together as part of a more holistic sense of spirituality that is just as available, if not more so, to children as it is to those schooled in particular spiritual practices or traditions. The dynamics that Sam and Rachel describe in play can also cross over into musical activity, broadening it away from simply singing fixed songs into an interactive process that draws on other engagements with the surrounding natural world:

> *Mark*: And so you mentioned sort of a summing up song or thing at the end … Does that then feel like an imposition or a break from the way you've been sort of feeling and relating to the world up to that point? Or is there a way of continuity there?
>
> *Rachel*: I think it's trying to bridge that space between everything that's out there and remembering that we're part of it and it's part of us. So I think it's kind of trying to pull the two parts together.
>
> *Mark*: And so that's invented in the moment, what you ended up singing?
>
> *Rachel*: The words are, the tune isn't … with my family groups it's kind of 'Here we go round the mulberry bush' is the tune.
>
> *Mark*: And what kind of words might they end up putting to that?
>
> *Rachel*: Oh, really depends what's been happening. So it might be we … curled up in the hammock, by the butterfly, or it might be … we create, we squidge the bird food. So the birds could be fed on a sunny day, because it's a sunny day today. (Rachel, interview with the author, 5 January 2022)

Play, here, is about an enjoyable engagement with the world around, and the creative ability to transform and create things in the moment. It is caught up in the core priorities of Forest Church, drawing on direct encounters with, and attentiveness to, the natural world. Music is something that can be played with and changed and, through this openness, songs connect outwards to what is happening around them, drawing them together, emphasizing human–nature interaction and commenting on that activity so that the group become aware again of what they have done and what they have encountered.

Pragmatic responses and considerations

As with many of the other groups, pragmatic considerations often shape the musical possibilities and directions as much as any more aspirational ideals that different groups might have as to the role and potential of different sonic practices. There can be debates or reflection in different groups as to the best way to shape their activity around a range of different issues. One obvious question when singing or praying together is whether you want to hand out bits of paper with some music or a liturgy on them, or whether that is something that might get in the way of the different practices that you are trying to encourage:

> If you don't produce a service sheet, you're not giving the group and the congregation who haven't been part of the preparation as much chance to participate. So if you give them a sheet, they can, you can, do responses, you know, calling songs backwards and forwards. If you don't give them a sheet, you're more likely to be able to tune in to the natural world, but you have to rely on responses that people have memorized. (Barbara, interview with the author, 8 October 2020)

The simple absence or presence of a sheet of paper shapes both the possibilities that are open for collective activity and the kinds of relationship that people might go about establishing with the world around them, directing attention and assisting in the performance of unfamiliar material. Likewise, weather and landscape play an important role in determining which kinds of activities are comfortable and natural to engage in:

> When you're standing around in the outside, you don't want to be standing and just singing very much, it doesn't feel quite right somehow, because often the weather isn't terribly amenable. And even if it's a sort of pleasant sunny day, then movement is a natural thing to do, I think when you're outside ... quite often they were processional for us ... we would

often process to a different part of the churchyard for another part of the ritual. (Barbara, interview with the author, 8 October 2020)

A natural question to ask when instigating different musical practices is whether or not to use some kind of instrumental or recorded support. When people are hesitant to sing, often a bit of support from elsewhere can help to bring them out or can make music more rewarding. Some groups have used different instruments to help with the singing, or to play on their own while different things are happening. Some even use CDs, amplifiers or recorded music, a practice that isn't without controversy. Is this kind of technology natural to the outdoor environment? Do people need it to come together in the space? Is it an unwelcome, potentially unnatural intrusion, or is it a crucial means of setting an atmosphere and bringing people together? Each of the groups I visited over the course of this research had different answers to these questions. One used an iPhone and hidden Bluetooth speakers to play atmospheric instrumental music during a time of outdoor Communion, one used a drum to support the sung activities of the group, one involved a capella singing, and one avoided musical practices altogether.

Drumming and rhythm

One sonic practice that is present across a number of Forest Church groups is that of drumming. Outside of a purely pragmatic sense that drumming is probably something that works easily in an outdoor environment, it was initially somewhat unclear to me what this practice was about – did drumming have a spiritual dimension to it? Did it help individuals to relate to the space around them? Were the associations with indigenous practices the key driver behind its incorporation into Forest Church rituals? As with many practices, there seemed to be a range of different ways of thinking about and experiencing drumming in an outdoor environment. In line with the overall orientation of his Park Church group – which takes inspiration from Forest Church but has its own distinct ethos and setting – Sam's understanding of the possibilities of percussion is somewhat pragmatic in nature:

> Well, I suppose it's participative, isn't it? You know, it's quite easy to get people to join in. And I think it's sort of low-tech as well, I suppose that's what a lot of people are looking for, when they're going outside ... it sort of feels like it fits with that outdoor environment and gets everyone, or anyone who wants to, kind of involved. (Sam, interview with the author, 10 January 2022)

As someone who has sought to engage more extensively with the practice, David Cole's understanding of his practices in and around New Forest Forest Church has a slightly more comprehensive and more explicitly spiritual rationale behind it:

> Drumming is something that's obviously found in a great deal of the indigenous spiritualities to bring a sense of the rhythm of nature into the physicality of the gathering. So that's really my focus when we were doing the drumming workshops to bring that reality of the rhythm and the vibration of the natural world into something that we can hear and engage with ... I think when you look at the indigenous culture, what drumming does within them is to stimulate the inner being, to stimulate that kind of spiritual context to ourselves ... there's research to suggest that the actual Earth is still vibrating. And obviously scientists ... will say this is the reverberation from the Big Bang ... so there is this sense of actually the Earth has a rhythm, has a vibration ... And with the sense of a heartbeat as well ... if the natural world has this resonance, and the Divine is in that, and there is this sense of this heartbeat of God, then they, we, you know, there is an overlap in those connections. (David, interview with the author, 19 February 2022)

Once again, sound is a means of tuning in and resonating with something divine that is present in the natural world, this time a sense of vibration. While Alison talked quite specifically of pagan spiritualities present in the UK, David's canvas opens up a little more broadly, to indigenous spiritualities as imagined on a global level. There is a sense of a cosmic spirituality that they are somehow tapping into, and that the group in the New Forest can, in turn, connect with through their own particular Christian tradition. Whether or not this understanding is truly reflective of indigenous people's own self understandings is, perhaps, a moot point. The theological lens that is being developed is implicitly universal in scope and application, rooted in a common physicality and presence in a shared created universe whose heartbeat transcends individual viewpoints and understandings. Even this kind of cosmic spirituality, however, has a pragmatic element to it, and David is highly attentive to the immediate dynamics and practicalities of the group:

> They all picked up sticks and stones. And we were sitting on a collection of logs and stumps and things. So we were all just kind of using whatever we had to bring a beat in. And it was to start with, we spent a bit of time just being out in nature, setting the rhythms, talking about birdsong as well, so connecting with the actual songs that are natural, and the music of nature ... I began to create the rhythm and everyone joined in. So there was a part of a collective rhythm making as well which, from our kind of subconscious psychology, connects us together as a group. It's a bit like collective contemplation, so sitting in silence in a group feels different to sitting in silence on your own, because of the collective energy. And I think that's similar to what happened when we were doing bringing a rhythm together, being together. (David, interview with the author, 19 February 2022)

Drumming together can be a matter of using natural materials, of being in an outdoor space and of connecting together through a group activity. Physical and interpersonal dimensions sit alongside those of a further-reaching spirituality, and it seems highly likely that the experiences of different group members will take place on different levels of experience, and even flit back and forth between them, some enjoying the activity for the fun that it offers, some experiencing enjoyment and connection in the immediate space around them, and others, sometimes, discovering a deeper sense of spirituality and encounter that taps into something beyond the immediacies of that shared space of activity.

Attentiveness and the avoidance of musical practices

Music is far from ubiquitous in Forest Church gatherings and, once I had established initial contact with particular individuals who were already known for their music, it was often hard to locate groups and individuals who felt that they really did engage in significant musical or sonic practices. For many groups, musical practice is occasional, low-key, ad hoc or even avoided altogether. Rachel, for example, while used to using music and song in a classroom situation, is wary of using it in outdoor environments:

> I really don't sing that much outside. And I think partly that's because, when we're trying to bore down into that connection with nature with the children, you don't want to put too much stuff in the way ... So we sing a song at the beginning, where we're talking about the space, about the weather, what we're going to be doing. At the end, we use the same song, but I make up verses that kind of recap what we've been doing. So kind of setting the scene of where we are in the place and what's happening in the place ... And so we'll talk about the things that we've noticed in nature, and put them in the song. I think that the reason we don't sing is because I'm wanting people to notice the sounds around them ... you're very aware when you're singing outside of the fact that there's that human tendency and temptation to control

and put your mark on a place. And when you're … talking a lot about listening to what God's got to speak to you through the fifth book of the Gospel, of creation around you, it's very hard to do that when you're imposing your humanness on it. (Rachel, interview with the author, 5 January 2022)

There is a wariness of imposing the humanity of participants on the natural environment, a wariness that comes together with an attentive orientation seeking to pick something up from the natural world around, either for its own sake or for a sense of the ways in which God might communicate through it. The music that does take place is developed around precisely the same orientation, incorporating things about the natural world into the brief moments of singing that happen precisely in order to *re*orientate participants back to that world and not simply on the human singing community.[3]

Another reason behind a reluctance to engage too much in music outdoors rests in the establishment of an intentional distinction from the forms and patterns of indoor worship, an abandonment of something that might lead to slipping into old habits or ways of thinking. Sam's experiences are grounded in a sense of this distinction as well as the more immediate practicalities of what feels comfortable and rewarding to engage in amid the particular outdoor space that his group uses for worship:

We read the Bruce Stanley book, the Forest Church book, and … he has this bit about, don't try and do indoor church outdoors … And so Park Church very much has ended up being about kind of trying to do worship that is indigenous to the space that we find ourselves … In terms of music in Park Church, we occasionally do songs with a ukulele or songs a capella, or we've done kind of hitting fallen logs and things to get a rhythm going or clapping or … but it's not a sort of regular part of what we do … and I think that's also partly to do with just the people that we've had come … I do remember a couple of them being like, 'Oh, we find it a bit embarrassing

when we sing, because ... the park is a very public space'
... We definitely wouldn't do a whole load of wordy songs,
because we don't want to be giving out loads of [bits of paper]
... Also, I do think that the acoustics of singing indoors are
kind of a little bit more rewarding. (Sam, interview with the
author, 10 January 2022)

The embarrassment factor adds an additional dimension for
Sam, since their group meets in a more public outdoor space in
a park. The implicit danger of being overheard by other people,
who are far from the intended listeners, is one that some are
more wary of than others; however, since Forest Church often
welcomes a range of individuals with a variety of different commitments and experiences, there is a need to ensure that a broad
coalition of individuals are comfortable with what is going on
in the space. An avoidance of indoor musical practices is not
just something that is done for the sake of it, but often out of a
sense that these practices, and the typical repertoire of Sunday
services, just aren't suited to the outdoor space, which demands
something different. Since there are no readily available musical
practices to turn to, this reinforces groups' move in a different
sonic direction.

In resonance with some of the literature on mindfulness discussed at the start of this chapter, often the alternative sonic
orientation revolves around attentiveness and silence. Indeed,
in my visits to Forest Church communities, this space for quiet
attentiveness to nature was common to every single one of
them. David Cole describes a pattern that I encountered in a
number of different Forest Church spaces:

I suppose that in our Forest Church, there is a sense that there
is a bigger element of silence, I think, than most others. And
perhaps listening, but not in an audible physical context. So
basically, the layout of our gatherings ... is that there is a
short introduction to the theme. And then people are given
about half an hour to go and just be in nature, on their own
or in ones or twos, or twos or threes. To kind of contemplate

on that thing … there's an introduction to the theme, half an
hour being out in nature, and then coming back and reflecting on the experience you've had. Our aim is to facilitate the
opportunity to experience something in and through nature.
So in that sense that half an hour is given over to silence and
listening at a deeper level, not just listening to the birds or the
wind, but actually listening to your own heart, listening for
the divine voice within the natural world. (David, interview
with the author, 19 February 2022)

Some shared conversation, reflections, liturgy or readings at
the start leads into the chance to go out into the surrounding
park, forest or garden, to be attentive to something particular
in that space, often following a particular suggestion from the
leader of the meeting. Participants then come back to share that
encounter in a conversational environment, sharing with the
others how something in the space around them led to a sense
of what God might be saying in and through it.

My own experiences of this time were slightly ambiguous in
nature. In almost all the spaces that I visited there was a mixture of human and natural sounds, and I was often aware of a
fine line between reflections on what God might be saying to me
and my own capabilities for constructing meaning in relation to
the different things that I noticed around me. Sometimes that
was an enjoyable practice, but mostly I had a sense of being
alone with my own thoughts, wondering about the different
prompts that we'd been given and conscious of the need to
go back to meet the group at a particular scheduled point in
time. I welcomed the democratizing potential of enabling different people to share what they'd experienced, and the chance
to be attentive to the natural things around me, but wondered
how close the theory and practice of these experiences really
matched up, or whether there was a hope for encounter that
went beyond the mundane experiences that were often possible
in the prescribed time and frame that were allocated to us.

Even ambiguous experiences can, however, be worthwhile
and worth holding on to. In an article on experiences of out-

door sound art, Joseph Browning highlights the value of 'de-escalat[ing] scholarly claims about the power of sound art and argu[ing], instead, for careful attention to its uncertainties and ambivalences' (2021, p. 188). He suggests that it is often not the ability to offer profoundly meaningful experiences that is important, but rather the surprise that, amid uncertain and ambiguous relationship, experiences of sound in nature are able to shift our awareness at all:

> ... it is here that we see the value of theorising sonic engagements with more-than-human sociality in terms of the 'attempt'. Rather than sound art's efficacy or power, guaranteed by the immanent relationality of sound and listening, it is instead sound art's contingency, animated by the uncertain relationality of sound and listening, which means it opens onto vast speculative worlds. (2021, p. 188)

We should not, perhaps, expect Forest Church to always offer profoundly transformational or intense experiences; rather, we should notice that, in attempting to reach towards the divine through nature, it provides some room for subtle shifts in awareness and attitude, sometimes offering something bigger, though often remaining much more low-key but nevertheless real and significant.

Spiritual intimacy and resonance with nature

While my own experiences were rarely particularly intense in nature, for some a sense of encounter goes much deeper than for others, and Annette's descriptions of her experiences at Forest Church YYC in Alberta, Canada, offer perhaps the strongest example in my fieldwork of nature as a space for deep internal transformation. The Forest Church in question is not connected to the Forest Church movement in the UK, nor is it explicitly set up as an expression of Christian faith. Rather, it emerged independently in connection with Annette's therapeutic engagement as a herbalist and in dialogue with her Catholic upbringing. It was an attempt to recover a spiritual dimension that was missing from her clients' experience of wellness and capacity to heal, with songs acting as a way to come together despite diverse religious backgrounds and perspectives.

The sonic practices of the group involve dwelling with individual songs for extended periods of time over the course of the year, and rewriting songs in an attentive relationship to one another and the natural world around them. This sometimes involves a sense that a sound or song has been given to them by the surrounding forest as they listen to it, and the journey involves a deepening sense of self, relationship with one another, and relationship with the Earth. It can involve mimicking natural sounds in a process of becoming one with nature, or producing sound with natural materials. This seems to involve a sense of inner wrestling as relationships deepen, and barriers to those relationships are brought down on a spiritual, internal and interpersonal level. Annette often framed the experience in terms evocative of a long-term therapeutic journey of intimacy with the Earth, acceptance of self, and intimacy with one another. Use of a songbook gives a stability of repertoire but also acts as a gateway into other sonic interactions, reworkings and experiences rather than as a goal in and of itself. Annette emphasizes the importance of resonance, of intimacy and of a spiritual connection with the natural world that connects with processes going on inside of individuals on a deep personal, spiritual and existential level:

> Our last Forest Church, we went into a dark forest ... It's very, very windy ... So everybody found their way through in mimicking that wind until we could find our resonance together, and with the wind, we spent about 45 minutes sounding, also moving as the wind, we became trees, first, in our movement of looking up and seeing them move. And then as we found our own spirit, or our wind, our own air, we stayed until we had some aspect of harmony ... at the beginning, there was a whole lot of 'Wow, I never imagined

that was possible'. Right, that sense of self lost, a sense of merging, people experienced awe for the first time, I was terrified by not knowing what was happening, but excited at the same time ... there's a lot of us moving kind of towards faith and trusting that something will be there, that something will move through us, that I'm not doing this alone ... We'll often ... start with a song ... and then we get to this big field where the wind has just rolled all the grass ... and we all just lay in that grass. So the words [are] an overall concept that's moved through our bodies. And by the time we're lying in that grass, those words are gone. They just got us there ... ultimately, for me, sound is resonance. And that's sort of its goal ... the resonance with the grass ... the sense of oneness ... Nature ends up moving into our relationships with each other. (Annette, interview with the author, 11 December 2020)

There are parallels to be drawn here with Barbara's sense of divine and natural encounter. In encountering nature, something else is opened up. Something more is to be found within and beyond it, whether that be God, spirit or something relational connecting to other members of the group. Later in our conversation, Annette described an encounter with a more explicitly therapeutic dimension to it, highlighting the way things that happen and resonate in the natural world have the potential to trigger parallel processes inside the people who experience them:

So I had an idea of just going back to where I was in childhood in this resonance ... Because there's not a trail, we have to kind of climb the mountain and we're overlooking the Columbia icefields ... [it's] guaranteed if sitting there for at least three hours, you will hear an avalanche. You'll hear the ice, you got to be patient and you got to pay attention and you have to have presence. And it was an amazing experience because not many people will get that resonance, because, well, even if you do, you don't want it because you're right in it. And so that was really one of our most profound experiences where

people felt things just break through in them and break off
... that's just probably my most profound sound experience
of, you know, going looking for the sound and waiting for
the sound and wondering what it's going to do inside of you.
(Annette, interview with the author, 11 December 2020)

Here we see the power of sound as a relational, embodied but
also spiritual, psychological and interpersonal phenomenon.
The cultivation of intense experiences and relationships over an
extended period of time in order to embark upon a shared journey of healing and discovery involves a degree of commitment
and involvement that many Forest Church groups might shy
away from. It is hard to reconcile these dynamics with a desire to
be open to outsiders and to create a low-pressure environment,
or with an ethos that is often based more upon the dynamic of
the parish where community and insider–outsider distinctions are
somewhat blurred. Nevertheless, the renegotiation of self and
other, human and nature that Annette describes is common in
some sense to many Forest Church communities, and Annette's
descriptions show the possibility of sonic and embodied practice
to involve deeper interpenetration of self and other than more
cautious practices might naturally seek out. The similarity and
contrast to the other practitioners who I interviewed highlight
both a shared sense of possibility in human–nature encounters,
and the way that these encounters are not simply given as a
universal mode of natural engagement but are constructed in
relation to the particular attitudes, frames and expectations of
each particular individual and group. Nature encounter is far
from a universal common category of experience, but one that
opens out to our particular imaginations and practices of what
is possible and what we aim to foster in that encounter.

Idealization and what comes after

The Covid-19 pandemic was, perhaps, a high point in the story of outdoor worship, with many groups inspired to try out outdoor rituals because of the risks of indoor services. During this time, the Church of England set up resource pages to assist groups wanting to transfer more of their services outside, while the warm summer weather provided the ideal opportunity for them to try this out without worrying about individuals' resilience to cold or rain. In ritual terms, Forest Church can be immensely appealing, offering a more radical shake-up of traditions of worship than is possible in church buildings alone. These positives are not, however, without possible dangers. While many of those engaged in Forest Church find it possible to encounter God in some way through the natural world, some who I spoke to were a little more cautious in their assessments of the movement, drawing attention to the ease with which humans sometimes idealize a nature-connected lifestyle and the potential hostility of natural environments as potential weaknesses in the movement's approach. A number of recent authors have pointed to similar themes in relation to recent moves towards pagan spiritualities. Anna Fisk, for example (2017), notes that, while modern pagan movements that seek to reconnect with nature are often seen as a positive move in the face of the global environmental crisis, there are more complex dynamics at work when constructing new religious practices in this area. Fisk notes the risks of imperialism, of appropriation from indigenous cultures, of romanticization, essentialization and the relocation of indigenous world views as an imagined part of a European past. Fisk doesn't suggest an abandonment of the movements but, rather, the need to be sensitive to these dynamics, to preserve a sense of otherness and to avoid pretending that a tradition is something that it is not. Terry Gifford likewise offers up potential critiques of pastoral idealizations of nature that are often bound up with the idea of escapist retreat, and a failure to question broader power structures underpinning regimes of land ownership and society as a whole. However, he also suggests that pastoral dynamics can offer political critique and resources for change, particularly if extended into a critical pastoral or post-pastoral mode. Such an approach would encompass both the awe and attention found in pastoral materials and a more nuanced understanding of what this entails, including an awareness of both creative and destructive natural processes, an understanding of the complex interplay between nature and culture, and a sense of our own responsibility amid patterns of human exploitation and injustice (Gifford, 1999, 2017).

A range of authors have suggested the possibility of holding together the darker side of the natural world with a desire for deeper relationship.[4] Kathryn Rountree suggests that, while it 'might be argued that other-than-humans such as tsunamis, mosquitoes, and cancer cells do not seem particularly well-disposed towards humans' (2012, p. 316), and while this might pose potential challenges when neo-pagans want to understand and experience these different beings in terms of kin, 'we could note that human kin are not always well-disposed towards one another either: human family relationships are often characterised by conflict, violence, and domineering or exploitative behaviour' (2012, p. 316). If the dark side of different relationships doesn't prevent us from using language of connection and relatedness in other areas of experience, then perhaps it shouldn't prevent us here either – rather, in all areas we need to stay aware of the complexity of the different relationships we are establishing, the dark along with the light, without letting the dark close off the possibility of working with or seeing the positive dimensions at work in different practices.

More than any of the other practices described in this book, Forest Church groups focus on a rebalancing of the human–nature relationship within the ritual experience itself. Nature and experience of nature often take precedence over and above pre-existing ritual frameworks. Although these rituals are not primarily directed at the climate crisis, they speak instead of a reconnection with nature amid a prevailing religious culture that many feel has become in some way alienated from natural engagement. While it should be clear from this chapter that Forest Church sonic practices are diverse and vary according to the priorities, beliefs and spiritualities of individuals and groups, music and sound participate in the renegotiation of relationships with other species and environments, sometimes awkwardly, sometimes profoundly. Direct encounter with plants and outdoor spaces means the role of human sound has to be reconceptualized in ways appropriate to these environments, and this can entail increasing intimacy with natural sound-worlds, a retreat from overbearing forms of human-produced sound, or allowing aspects of the surrounding world to take on an increasingly prominent role in forming the music that humans produce. The distinction between music and sound is one that becomes increasingly blurred as the sounds of trees and animals take on meaning of their own, and become bound up with new ways of imagining the spiritual relationships between divine, human and non-human beings.

In the context of this project, Forest Church experiences help to demonstrate the kind of sonic reworkings that can take place when groups seek to base their ritual practices on direct encounter with nature rather than simply bring nature into their existing ritual practices. They can help make visible some of the implicit norms and understandings that govern sonic and ritual activity in an indoor environment, and the way in which it can be helpful to draw these into question. Nature spiritualities are not without power, and the literature shows that they can shift and nudge people in positive directions bit by bit over time. These are complex and varied spaces and, as we have seen, a critical understanding can open up both space to acknowledge their potential pitfalls and the chance to move beyond them.

Notes

1 Laurel Zwissler (2011) discusses earlier progressive Christian borrowings from neo-pagan traditions, including a discussion of what it means for different traditions to borrow from one another and the

legitimacy of such borrowing in relation to dynamics of power and intercultural encounter.

2 Beyond the building of connections with pagan or indigenous practices, just as with Christian Climate Action, Forest Church musical practices intersect and overlap with musical practices being developed outside of distinctively religious or Christian environments. In conversation, Barbara drew attention to a range of other projects, such as the Lost Words project – a book and album that developed out of the disappearance of common nature words from children's dictionaries – and Sam Lee's activities engaging in music in outdoor spaces alongside nightingales. There is a sense of nature and the environment as places that are a common concern for a range of people both within and outside of Christian faith communities.

3 Not all sounds are simply those of nature, and often there are a range of sound sources in the immediate area, some technological, some human, alongside those of plants, landscapes and animals:

> The sounds are interesting, because we are in the flight path for Reagan National Airport in DC. And that actually is really cool. Because you have interspersed with the sounds of the crackling of the sticks under foot, and the rustling of the leaves, and sometimes the lap of the water against the riverbank, because often we end up on the banks of the Potomac River. So you have this beautiful riverscape. And on the other side, it's Washington DC. So you have this real juxtaposition visually, but also audially, I don't know if that's the right word, but because you hear like the sounds of things that are made by man, but you also hear the sounds of things that are made by God. (Pete, interview with the author, 20 December 2021)

4 David Waldron and Janice Newton (2012) point in a similar direction, while Vanessa Sage (2009) highlights the origins of contemporary pagan movements in nineteenth-century romanticism. Michael York (1999) offers helpful context on the constructed nature of contemporary paganism and on matters of appropriation (2001).

Gauging the Climate

The five snapshots that we've explored over the course of the different chapters are very much just that. They showcase particular groups trying out particular activities in particular places and particular moments. There is relatively little indication in the research I have carried out as to whether any of these are likely to be adopted on a broader scale, or whether they offer sustainable forms of ecological spirituality that can shape communities in an ongoing way. These limitations, however, do not mean that there is nothing to take away from this research journey, and in the remaining pages of this book I want to explore what broader conclusions we might be able to draw and what they might mean for future ritual practices in different communities.

At the very simplest of levels, it is clear, first of all, that musical responses are one of the means that Christian individuals and groups are beginning to turn to in navigating the current global situation. This dimension sits alongside more practical actions that people take, sometimes as a direct attempt to spur practical action, but sometimes as part of a broader project of integrating practices of faith and feeling with the changing world around them. For some, turning to music seems to be an instinctive response that provides numerous potential avenues to explore.

There are many different reasons for turning to music as a means for engaging with ecological themes. For those already active in different musical contexts, the growing urgency of the climate crisis and ecological alienation is increasingly something that needs to be grappled with and, as such, often demands some kind of attention, regardless of whether it is an easy fit within established paradigms. Importantly, music is a realm where this kind of creativity is possible and permitted; where someone who feels the impulse to write or perform something can try this out and test its impact. Because of wider institutional recognition of the importance of the climate crisis, there are an increasing number of possible forums in which this kind of creativity may be welcomed, and where an individual may not find themselves completely alone in exploring these themes through music, instead finding themselves allied with a range of other individuals who can see the value of this endeavour and who may be willing to come on at least a part of the journey alongside the musician.

Not all of those involved in the different initiatives would understand themselves as musicians, and for these individuals it is the different possibilities afforded by music and sound that come to the fore in relation to the different initiatives and projects that they are interested in exploring. Music can be a way of grabbing attention, of managing emotional journeys, of relating to the surrounding world. Moreover, there are a range of different entry points into sonic creativity, whether through adapting lyrics in accordance with given structures and tunes, performing in novel locations and situations, improvising together with others or working together with the dynamics of percussion or natural materials. Outside of the high-pressure world of on-stage musical performance, many of these activities can be tried out in participatory groups and communities, with the primary indication of success or failure consisting in the effect on the different people present, rather than the complexity, perfection or virtuosity of the musical product in itself. This all means that music is a multi-dimensional and accessible medium that offers rich possibilities to be uncovered. The nature of the possibilities available is highly specific to particular situations, but music's flexibility as a medium means that it is adaptable to a range of use-cases and scenarios.

Critical tensions

The diversity of different communities and motivations, together with the exploratory nature of much of the creativity described over the previous chapters, means that there is currently no single common musical route that provides a template for different individuals and groups to adopt. Instead, the connections between environmental change and the musical expression of Christian faith are many and varied, and they draw in a range of other concerns that are more or less important within different strands of Christian belief and practice. Some expressions are more radical than others, depending on the extent to which those involved believe existing traditions need reworking in order to grapple with the issues being encountered. This is completely to be expected, but immediately raises the question of how to understand this diversity of practice.

We can approach this question from a number of different angles, and musical dimensions can, in part, be seen as a manifestation of broader dynamics and tensions that play out on a range of different levels and areas of interest when it comes to matters of the climate and our relationships with the world around us. The Australian philosopher and ecofeminist Val Plumwood has pointed to the range of tensions that can be present in spiritual responses to nature as much as in any realm of human grappling with environmental relationships. Plumwood suggests that:

> ... far from being in some generalised and indiscriminate sense 'the answer' to our difficulties in coming to terms with nature, [spirituality] has many of the same ambiguities and potentials to foster better or worse relationships with nature as other kinds or theories and practices. The problems of reason/nature and mind/body dualism, human-centredness and self-enclosure, remoteness and use/respect dualism arise for spirituality in much the same way as for areas like ethics. (2002, p. 218)

For Plumwood, there are a range of different potentials that can be taken advantage of and pitfalls to be avoided when thinking through issues of nature and ecology in relation to spirituality, or indeed anything else. Spiritual responses to the climate crisis can reinforce unhelpful dualisms and divisions in our thinking or reinforce problematic hierarchies and relationships as much as any other medium for negotiating these relationships. Different expressions are likely to handle particular tensions better than others, and this needs to be kept in mind when navigating between them. Ecological criticism, of the kind that Plumwood engages in, often focuses on the explicit or implicit way in which the relationship between humans and the surrounding world is imagined and framed by different communities and in different documents or artistic creations. Does a particular avenue serve to over-idealize or aestheticize an artificially constructed ideal of nature? Is nature viewed as distinct and separate from humans or as bound up with them from the start? Is the world imagined in an overly instrumental or dominating fashion? Have we established an ontology that really does justice to the range of beings out there in the world and to our role in this broader ecology? (Latour, 2004; Morton, 2007; Plumwood, 1993, 2002)

The participants in the different projects that I have described are not naive on these kinds of issues. Many of these questions are already being asked by the communities and individuals interviewed over the course of this research, and individuals are often aware that they are still negotiating these questions and have yet to approach the endpoint they might ultimately hope for. There are easily visible locations for optimistic or sceptical points of view to latch on to in almost all of the movements, and each individual development taken by itself can very easily be made the subject of critique. We can criticize the evangelicals for buying into the logic of an ecologically destructive market as much as Forest Churches for a naive back-to-nature spirituality, or we can praise the evangelicals for their pragmatic use of music to bring an issue to the fore and the Forest Churches for their willingness to bring the major structures of Christian worship into question.

It is important to examine the practices of these communities through, at a minimum, a double lens. An ecocritical lens must necessarily, in this context, sit alongside an acknowledgement of the different theological and spiritual priorities that serve to guide the different groups. The movement to innovate – and to innovate in a particular direction – in response to changing ecological dynamics takes place in relation to the specific affordances that a particular setting allows and enables. The priority that is placed upon an authentic encounter with God in and through contemporary worship music (Porter, 2020, pp. 71–92), for example, will allow certain ecological priorities to come to the fore while pushing others into the background. Those broader priorities may end up being reworked within some groups over the longer term as they grapple with issues around them; however, many will seek, first of all, to build upon the affordances that exist within a tradition, and to explore this potential before seeking to stray further into the unknown territory beyond. In contrast to an evangelical pragmatism, for example, the extent to which Forest Church communities foreground nature and consider alliances with forms of spirituality from other traditions can provoke a level of suspicion from those with a concern for maintaining certain varieties of Christian orthodoxy (Nita, 2016, p. 159). The choice to undertake more fundamental reworkings of traditional structures ultimately results in the addressing of their offerings to a different constituency, which itself faces up to immediate limitations precisely because of its more radical orientation to faith and spirituality.

Forrest Clingerman highlights the diverse range of narratives that Christians and other religious practitioners can hook into, suggesting the way in which this area of practice is contested between different visions of the world around us. According to Clingerman:

The competition between narratives can be fierce. Some religious narratives emphasize environmental care, while others accuse environmentalism of being an idolatrous worship of Gaia. Some stories promote asceticism, others moderation, and still others material prosperity. And, of course, some say that only God has the power to change the climate, while others say humans are capable of being drivers of global change. (2015, p. 346)

Similar dynamics play out on a musical as much as on a theological level, and these tensions are something that musicians will need to navigate in positioning themselves in this renegotiation between climate crisis and faith tradition. Music participates in a range of familiar tensions regarding faith and climate change, which manifest themselves in debates surrounding theological approaches and appropriate Christian responses. But it does more than this: music provides a space for the renegotiation of attitudes and relationships, for reworking or solidifying existing traditions, for the balancing of power and priorities and for the engagement of both emotion and the physical world. It has the potential both to create and to address feelings of alienation between belief and environment, and as such it should be clear that Christian musical activity is neither a meaningless distraction from the real work of concrete climate action nor something that provides a direct solution to human-induced climate change. Rather, it is bound up as part of the same world of concern as a field of action participating with, and alongside, a range of others, and with its own contribution to make as part of this wider landscape.

The necessity of diversity

Tension and competition are not the only way of envisaging this kind of diversity, however. In his contribution to *The Oxford Handbook of Ecocriticism*, Richard Kerridge draws attention to the range of different tasks that need to be accomplished in relation to the climate crisis, and the potential for different literary genres to help with different tasks, albeit in a highly provisional and continually changing manner. Drawing on the work of James Lovelock, he highlights the need for both urgent technological solutions and long-term shifts in social and personal values:

> For literary ecocritics, Lovelock's proposal suggests that we might allocate different tasks to different literary forms and genres. Prompted by Lovelock, we might say that we need all the different literary forms to do different jobs: realistic novels and lyric poems, action thrillers, realist and Brechtian theater, avant-garde forms in the Modernist tradition ... If the fundamental aim of literary ecocriticism is that environmental care should become stronger and more pervasive throughout literary culture, ecocritics will not be looking for a single form of literature that meets all the criteria at once; nor will they search only for a small number of new forms or genres specially adapted to environmental priorities. Rather, they will want to address all these various needs and audiences, and to bring environmentalism into all the influential forms of literature ... the emergency does not give us leeway to entrench ourselves in our taste for one literary genre or another. Ecocritics must continually look at the need to reach a variety of audiences, and to address different emotional reactions to the crisis, and to accommodate conflicting tactics. The crisis makes us twist and turn. (Kerridge, 2014, pp. 369–70)

Scholars in different areas make similar claims. Duncan Green, for example, in his work on the nature of change, suggests that organizations often do best to think in terms of a portfolio of experiments, which compete and evolve over the course of time. Diversity leads to innovation and resilience, with some attempts succeeding and others dying out when they fail to take hold (2016, pp. 14, 251). For Green, diversity can be a pragmatic strategy to cope in a complex world where there is little guarantee that a particular approach is likely to bear fruit on its own. In a similar manner, Laura Hartman focuses on climate change as a problem that is both wicked and wild in nature – which has multiple and complex causes, and defies our attempts to solve it in any kind of straightforward manner:

Climate change, and other environmental issues, are mammoth imbroglios – haunting, overwhelming, at once spiritual and physical, political and psychological, multifaceted and tricky … scholars should resist any attempt to find one golden solution [and] the problem should remain wicked – or wild – lest others falsely believe it has been solved … The overwhelming dimensions of climate change (its global scope, its mind-stretching time scale, its vastly varied on-the-ground impact) must not be over-simplified, abstracted, or tamed. We humans must recognize our smallness, our partiality, and the incompetence of our theories and tame, discipline-based approaches … If the problem is wild in its essence, this calls for responses that address uncertainty and risk as well as the generative and creative dimension of such ambiguity … Such a wild, wicked problem invites collaboration and collective problem solving. Mammoth imbroglios like climate change … stem from multiple entangled causes, they remind problem solvers of all sorts that there is an urgent need for cooperation across lines of difference. Looking at the problem through multiple lenses creates a prismatic array of definitions of the problem, and hence of solutions … Climate change is deeply multiple: multiple causes, multiple definitions, multiple analyses, and multiple solutions make up its existence in the world. (Hartman, 2017, pp. 91–5)

The need to pursue multiple avenues in navigating the current climate crisis arises not simply because greater numbers often help to make a greater impact, but because the causes of our current predicament are multiple and complex in nature and, therefore, require this by their very nature. We should not be surprised by the range of initiatives present in this book, nor should we attempt to narrow them down to a single approach that is better or more helpful than the alternatives. The climate crisis does not allow for this, and we need to search for a way of holding together this diversity and harnessing it in a way that works in the midst of multiplicity and difference.

A musico-ecological-spiritual constellation

Following both my own previous suggestion that different traditions of Christian musical activity operate within a dynamic
ecology in which each serves to open up dimensions of world-
relation less easily foregrounded within other alternatives
(Porter, 2020), and the pluralist approach of David Ingram
(2010), who finds value in the diverse contributions of different
musical forms and approaches to issues of environment and
ecology, I want to suggest that current moments of innovation
that are occurring across a broad range of groups might be
better understood not in isolation from one another as stand-
alone attempts to fashion a single guiding form of Christian
ecological relationship through music but rather as a constellation that is only able to adequately address itself properly
to the full range of necessary ecological engagement precisely
through its multiplicity and disjuncture (Krauß, 2011, p. 443).

Timothy Morton has contrasted the almost-disappearance
of nature that happens as deep ecological approaches to the
world bring the human and non-human close together with the
prejudicial concepts we may be left with if we seek to erase that
distance too quickly: 'Hanging out in the distance,' he suggests,
'may be the surest way of relating to the nonhuman' (2007,
pp. 204–5). The presence of contrasting musical expressions
can play an important role in balancing these kinds of tensions, not just in relation to closeness/distance to nature but in
response to issues of pragmatism/idealism, grief/hope, divine/
human action. The different musical responses here are not necessarily to be understood as direct reactions to one another;
rather, they emerge out of traditions of faith and music that
have arisen in this way and that each come into dialogue with
the same current environmental crisis already in both agonis-
tic and irenic relation. As such, they form part of a dialogue
between different forms of spiritual-musical world-relation and
a common ecological challenge. The constellation that they
are beginning to form is very much a work in progress, and
many of the groups in question currently operate without any

broader awareness of the activities being undertaken outside their own immediate circles. However, I suggest both that it will be through the further development of a dynamic ecology of different expressions that serve to critique and complement one another through their ability to address particular aspects of human–ecological relations with greater or lesser adequacy that innovation is most likely to proceed and that it is through this kind of process that an adequate response to the ecological crisis might ultimately emerge.

This notion of constellation can be elaborated further by reference to the work, in the 1990s, of Etienne Wenger. In developing his ideas surrounding communities of practice, Wenger found it necessary to explore situations in which different groups are not knit together into a coherent community but relate to one another in a more distant manner. Wenger suggests that:

> Some configurations are too far removed from the scope of engagement of participants, too broad, too diverse, or too diffuse to be usefully treated as single communities of practice. This is true not only of very large configurations (the global economy, speakers of a language, a city, a social movement) but also of some smaller ones (a factory, an office, or a school). Whereas treating such configurations as single communities of practice would gloss over the discontinuities that are integral to their very structure, they can profitably be viewed as constellations of interconnected practices. (Wenger, 1998, pp. 126–7)

The idea is an attempt to hold together both relatedness and discontinuity, and Wenger further elaborates what this relatedness can look like:

> The term constellation refers to a grouping of stellar objects that are seen as a configuration even though they may not be particularly close to one another, of the same kind, or of the same size. A constellation is a particular way of seeing them as

related, one that depends on the perspective one adopts. In the same way, there are many different reasons that some communities of practice may be seen as forming a constellation, by the people involved or by an observer. These include: (1) sharing historical roots (2) having related enterprises (3) serving a cause or belonging to an institution (4) facing similar conditions (5) having members in common (6) sharing artifacts (7) having geographical relations of proximity or interaction (8) having overlapping styles or discourses (9) competing for the same resources. All these relations can create continuities that define broader configurations than a single community of practice. A given community of practice can be part of any number of constellations. (1998, pp. 127–8)

It can easily be argued that the different groups described in the preceding chapters have shared roots in Christian faith traditions, and have related enterprises or causes in addressing the current climate crisis. Wenger goes further than this, however, in suggesting particular ways in which the continuity within a constellation can be mediated by different objects and practices. He suggests the existence of boundary objects and brokering; of boundary practices, overlaps and peripheries; shared elements of style that are copied, borrowed and imitated between communities; and discourses that travel across boundaries and serve to coordinate and reconcile different perspectives.

It is, perhaps, the existence of such mediations that moves a constellation from a relatively arbitrary collection to something that, despite discontinuities, has a sense of relatedness that goes beyond the choice to look from a particular perspective or angle. This, I want to suggest, is one of the crucial tensions within the pages of this book. In describing these different activities, I am, to a certain extent, attempting an act of brokering, and engaging in coordinating discourse that helps to mediate between the different perspectives and practices. This goes as much for the text here as for the different conversations I had over the course of the research, describing different groups and individuals to one another during my various interviews, and waiting to see how my different conversation partners might react to that awareness of other conversations and other communities.

Some places of overlap and brokering exist, to a certain extent, already. An easy example would be the way in which some of the *Doxecology* songs have been shared and appreciated in the Christian Climate Action WhatsApp discussion group, or the involvement of musicians in both the songwriting for that album and in outdoor worship practices. The benefits of further mediating and brokering relationships are, I think, relatively clear, precisely for the ability to coordinate action in different areas and in order to expand the horizons of different individuals and communities. Shared environmental concern means that some of these naturally develop anyway, but it is important to highlight the ways they can emerge through shared media, the migration of practices, the movement of individuals, and through conversations. There are numerous ways for these relationships to be strengthened in future, many of which are able to emerge precisely on the smaller-scale level of individual migration and sharing, without an absolute necessity for the kind of large-scale structural coordination that can be unwieldy to build up and maintain.

What do the different projects do?

How, then, do the different projects we've seen over the course of these chapters fit into a constellation? What are their different roles and what is their current relationship to one another?

The climate-album projects of Chapter 1 operate on a couple of different levels. On a grassroots level, we see a range of different individuals with different priorities and motivations being drawn into projects that somehow touch a point of commonality between them. On an intra-organizational level, we see individuals grappling with the limits of traditions and with processes of continuity and change, figuring out the relationship between the pragmatic and the ideal. On the next level up, we see album products that enter into circulation and reach sometimes unexpected constituencies and interest groups, pro-viding them with resources that meet a recognized need in the current moment that is unmet from anywhere else. These are albums that find their role in an existing scene, but one that is changing and in which they seek to mediate and shape that process of change in a way that keeps everyone on side but moves them step by step into a different way of thinking and imagining the world around them.

The activist musicking of Chapter 2 inhabits a very different world, one that seeks to discomfort and disrupt and addresses itself to multiple audiences. In activism, music both nour-ishes the spirituality and emotional life of different groups of protestors and reaches out beyond them as their efforts are addressed to wider publics on the streets around the country. The projects here are shaped by pragmatic considerations as much as those of the climate-album groups, but those consider-ations are somewhat different as a result of the different context in which they take place. Grassroots creativity is not filtered through an attempt to create a sellable product, but becomes visible as a wider group feels comfortable in taking it up, join-ing in or making it visible in particular, often fleeting, moments.

The song festivals described in Chapter 3 represent the inte-gration of music into a wider corporate agenda for action. They are acts of organizational communication that reach beyond the boundaries of the organization but that exist as part of an attempt to bring a multi-national Christian body of people together in a common project incorporating organizational change, practical projects and a reshaping of spirituality. As much as the preceding two projects, there is a great deal of internal diversity in Chapter 3, with different projects reflecting different situations and priorities; but they are all shaped, in this case, by a common reference point in relation to an influ-ential papal encyclical.

The requiem projects of Chapter 4 have multiple roles, as projects to raise awareness, as therapeutic performances and as attempts to pioneer particular emotional narratives. They are loaded with a certain degree of uncertainty and ambiguity but find their role precisely in a future-orientated space where clarity is sometimes hard to come by and where a grappling with multiple possible narratives and different ways of seeing the current situation can be exactly what is needed in order to find a proper emotional orientation.

Finally, the Forest Church groups in Chapter 5 attempt to rework human–nature relationships in a very direct manner as music and sound provide a medium for encounter between people and space. The projects here seek a certain level of break with traditions, with the move outdoors detaching groups from familiar reference points and ways of interacting in buildings and sheltered spaces. They are not wholly traditionless, however, and seek to establish new points of reference, broadening their canvas to take in a range of inspirations both within and beyond Christianity, shaping existing affinities and experiences with nature in a way that might have power where many feel alienated from that world around them.

In characterizing the different projects in this way, I hope to show that each has a range of characteristics that are specific to itself, performing a role that can never be performed in the same way by any of the other projects or groups. Indeed, while there are points of commonality between the projects, in most cases acquiring the characteristics of one of the others would involve giving up the particular (and important) function that it performs. The different groups may well have something to learn from the priorities of one another; indeed, I want to suggest that they have a lot to learn in these encounters, but they can never be completely collapsed into a unified whole without erasing their ability to address a range of different priorities and situations in ways appropriate to their own particular requirements. If constellations entail brokering and boundary practices, this will inevitably involve friction, negotiation and transformation as that boundary-crossing takes place. That process of mutual inspiration, critique and sharing of knowledge and resources is crucial but far from straightforward. While the internal logic of each group works in the particular situation that it inhabits, the potential of this to become meaningful for other situations is dependent on adaptation and dialogue that understands the differences of the contexts and is ready to figure out what that means for the potential to work together or learn from one another in charting a path forward to an unknown future.

What different themes can we trace?

There is potential for some of that dialogue to build up through an awareness of common themes that we can already trace across the different chapters, and I want to suggest six different themes that can help us to understand some of the common considerations that arise across a range of different environments:

The first theme that I want to trace across the different chapters is that of coalition. A completely unified perspective is hard to find in any of the projects described in the book. Rather, each project or movement brings together diverse individuals and perspectives around a particular enterprise where they are able to invest alongside one another and contribute something that may well come from a different direction than, or be in a certain amount of tension with, other contributions. This is far from unique to these projects – similar mixtures of sameness and difference are to be found almost anywhere we look – but different institutions attempt to manage this kind of tension in a variety of different manners, and it is important to note how many opportunities there are in the individual projects for the diversity of the coalitions to become visible at different moments and in different ways.

The second theme is that of creativity. This theme was written into my research here from the start, since I have sought to document emerging initiatives and projects rather than institutions that continue to deploy old resources in new situations. Nevertheless, it is striking that much of the creativity documented here involves a degree of novelty that goes beyond simply the creation of a new repertoire according to established patterns. Rather, much of it involves an exercise of the imagination and an attempt to stretch out into new spaces and figure out what might be appropriate within them. It involves processes of trial and error, the gathering of inspiration from a range of different directions, and a level of strategic thought as to what might happen in the process of trying something new. Creativity here is profoundly relational in nature, and emerges precisely out of relationships with situations and communities in a way that contrasts with traditional depictions of artistic individuality. It is notable that this enables both musicians and those who wouldn't describe themselves in that way to enter into the creative process – something that is crucial in bringing a broader constituency on board to act in the face of the ecological crisis.

The third is closely related, and has to do with the grassroots and often somewhat provisional nature of the different initiatives that I have sought to document. Much of the creativity that I have tried to document is creativity 'from below', which offers a range of individuals space to try out something that may or may not endure beyond its initial moment of realization, but which provides something for a particular moment or situation and draws upon the initiative available within that context. This allows a close connection between creativity and community, and can also involve a degree of empowerment when the production of musical repertoire is taken out of the hands of professionals acting at a distance and put into the hands of those who need it for particular purposes.

The fourth theme is that of pragmatism. The practical constraints encountered vary a great deal across the different groups. Some are more conscious of the potential reactions of a different audience or congregation members; others focus more on the ability to engage easily in certain kinds of musical/sonic activity in particular environments where particular constraints are present on the kinds of resources you can easily draw on. Some think about the journeys that individuals are likely to be willing to embark upon and the steps that can easily be taken, and some think simply about the easiest way to get a message out to the largest constituency. None of the groups are purely pragmatic in orientation, and they all seek to embody particular ideals and convictions, without which their activity would become largely flat and meaningless. Nevertheless, each is very aware, in order to achieve their goals, that idealism will inevitably end up tempered by the realities of the situations in which they find themselves. Often this takes place without too much of a struggle, since the situation is, to a certain extent, a self-chosen one, each group going about its project in a setting that it has already decided it wishes to engage with.

The fifth is the importance of community and relationship. This is already heavily implied in some of the other themes listed so far, but is important enough to warrant a separate appearance on the list. Much of the activity that I have documented across the different chapters happens together with other people, and it happens for particular people. The needs of community can be an important spark to processes of creativity and, where that community is immediately accessible, the opportunity to create something that they will take up and use or engage with can offer a very immediate sense of reward that you have done something of value and importance. The climate crisis is one that is largely impossible to engage with on a purely individual level, but where global publics and structures can also seem too distant and alienating for action to take place. The mediating level of community is one that can offer a greater degree of significance without the overwhelming sense of disempowerment that comes with awareness of structures well beyond your own individual capabilities.

The final theme is the transcendence of boundaries. This is most easily visible in the inter-religious dimensions of Chapter 2, but it is present elsewhere too: the boundary between human and non-human in Chapter 5 or sacred and secular formats in Chapter 4. The climate crisis, as a common crisis, involves the establishment of new partnerships in a way that often problematizes existing boundaries and distinctions between groups and constituencies. In many situations there is the potential for this kind of transcendence to create a degree of angst around the maintenance of group identities or distinctives; however, a shared interest has the potential to relativize some of these boundaries and push them into the background. This is by no means automatic or universal, however; at the very least, a shared challenge provides the chance to rework and reassess who is working together and how in the light of the changing dynamics around us.

Inspiration and dissatisfaction

What can we take away from any of this? And what might it mean for the future we all face? I may not be able to make too many predictions, but perhaps there are lessons to be learnt nevertheless.

The diversity of the different projects and the different strategies that they adopt is something I have found highly inspiring. There is room for the reworking of emotional journeys, for forms of visible, public action, for crossing over into different faiths and traditions and for reworking the sonic interactions that we engage in with the world around us. I am convinced that there is something to be learnt from the diversity of responses that different groups are offering. There are some commonalities between different avenues but there are also significant differences in focus, direction and imagination. While this diversity is something that I have personally found incredibly important, my description of the projects in terms of constellation rather than in terms of something more closely related is, at least in part, a function of the fact that different groups were not always particularly aware of what the others were doing, a factor that could also serve to limit their own imagination of what possibilities might be available. To appreciate the diversity of recent musical activity, and in offering a portrait of some of that wider diversity, I have tried to nourish a broad imagination of what might yet be possible, and of the benefits and limitations of different approaches. Just look at what's happening out there, see the ways that different individuals have thought through different situations, and perhaps in learning from one another's experiences something might be stirred in our imagination for the future. As the same time, each of the different movements I've described has the potential to act as its own critique of the others, highlighting the things they are missing as well as the things that they get right. And that mutual critique is as important as the potential of each project to offer a certain amount of inspiration.

Indeed, I want to emphasize again that the further development of an ecology of different responses is crucial. Christian musical traditions rarely remain completely in their own isolated boxes, but cross over to be found in unexpected places as traditions evolve, as mainstreams change and as new forms of church emerge and fade. The appearance, evolution and dissemination of contemporary worship music is one of the clearest examples of the way in which new forms of expression have both appeared and made their way into a broad range of traditions far beyond the charismatic evangelical context with which they are most often associated. It is also a clear example of the way in which observing musical traditions crosses over far beyond the musical world, to include market dynamics, theologies of divine presence, political allegiance and generational change. Music has formed the focus over the course of this book, but the dynamics we observe in and through music ultimately point to a much broader range of themes that will be relevant in broader grapplings with Christian communities in a whole range of different areas. Music has a way of drawing attention to itself, of making something visible, but what it enacts has a way of drawing together a much deeper range of issues of crucial importance to individuals and communities.

Alongside inspiration I need to trace a parallel sense of dissatisfaction that would sometimes bubble up as I was encountering and writing about different initiatives. If we bring the kinds of projects described here into dialogue with developments in environmental sound art, inter-species musicking or ecomusicological work on indigenous musical communities, there is the potential for a great deal of Christian musical activity to look somewhat tame in comparison. Even at its most radical, it does not in general aim to offer something deeply unfamiliar and, to this extent, Christianity and the demands of Christian community seem to act as something of a conservative force. This can be interpreted in negative ways, as something holding us back from a more radical future that is needed, but also in positive ways in its ability to reinforce a grounding in existing communities and a connection to longer traditions of meaning-making and togetherness. In writing about these projects, part of me was often hoping for something that more radically unsettled our existing models of music in human community, something that pushed me to surprising places and ways of feeling the world around me. Many of the encounters I had over the course of the research offered ways of seeing, thinking and feeling that were new to me in some way, which I hadn't encountered before and where I revelled in that surprise. As I went through, however, I sometimes wondered whether any of this is really enough, whether it pushes people to radical enough change and whether it unsettles enough our existing ways of living in, interacting with and experiencing the world around us. Such concerns fall into the same pattern of tension already seen between the different projects, with more radical shake-ups needing to sit alongside more pragmatic measures, but I am left both encouraged to see what is out there and wondering what more might be possible.

Directions for the future

Personal inspiration or discontent does not, of course, automatically translate into a helpful measure of impact, and in bringing this book to a close it is perhaps important to emphasize that while there is much that we know, there is also much that we don't know and will need to keep exploring in order to better understand. In talking with individuals who are more on the creative than the receptive side of these movements, I have relatively little perspective here on the actual effects of what they have been doing in relation to a broader population or broader social structures. What would it mean for them to be effective? Is that even an appropriate category to think in for most of these projects? We certainly have little glimpses into specific hopes or moments of evaluation through the course of the different chapters, but within the context of this research they are simply that – little glimpses. Tracing the impact of the initiatives is a difficult task – do they change the world views of those who encounter them? Do changes in world view lead to lifestyle changes or changing modes of encounter with the world around? Are there broader structural changes that result? Do we see them feeding into changing political and economic dynamics? The projects are all aiming for different kinds of impact, and to evaluate them according to a common set of criteria would perhaps be unfair to try. Likewise, none of the projects operates in isolation, and the role that they play is part of broader initiatives and life in community where the role of music may sometimes be to accompany other changes in an appropriate way rather than to bring about those changes on its own. These initiatives are about wider patterns of adaptation and action as climate change and the broader ecological crisis filter down into different aspects of our common lives, and they demonstrate the participation of music and ritual in this broader process, without necessarily being able to pinpoint a specific kind of impact or consequence that results in relation to that musical activity as opposed to the other patterns and actions that it participates in alongside. At the very least, we see the way that engagement in musical creativity enables a space for thinking through some of the issues at stake, for processing them alone and in community, and for doing so in a way that both draws on and feeds into wider cultural changes and patterns of development. What the consequences of this ultimately look like is a question deserving of further exploration as we are caught up in a moment of possible but uncertain change.

The practices described in this book do not all rest on a sense of hope. Some do not orientate themselves to the future, some centre on loss and others are deliberately cautious about voicing a hope they are not sure they can really believe in. Some, in other words, might be suitable accompaniments even in a world that remains irretrievably broken, while others will seize their moment precisely as action gains momentum and something begins to be done. Ultimately, the stirrings that I have traced here will need to be part of much larger-scale transformations across all levels and areas of society if they are truly to be seen as part of an adequate grappling with the issues we face. The diversity of responses is something of a microcosm for a broader range of initiatives that require diversity and multiplicity in order to face the fact that almost every element of our lives and communities needs to change in some way to reorientate itself for a different future. This far-reaching change is the result of long-term failures at proper ecological orientation, together with the pervasive nature of problematic ecological practices on micro and macro levels. Something is happening, but it is not yet enough. I hope that there will be more, not just because we need it in order to survive but because the ecological vision is ultimately a truthful one, and one that I find inspiring.

Bibliography

Allen, Aaron S. and Kevin Dawe (eds), 2016, *Current Directions in Eco-musicology: Music, Culture, Nature*, New York: Routledge.

Arnold, Jonathan, 2016, *Sacred Music in Secular Society*, Abingdon: Routledge.

Ashley, Peter, 2007, 'Toward an understanding and definition of wilderness spirituality', *Australian Geographer* 38.1, pp. 53–69.

Băncilă, Maria Yvonne and Iulia Alexandra Chertes, 2022, 'Peter Reulein's Laudato Si' Oratorio: women's voices endorsing ecotheological tenets', in Nadja Furlan Stante, Maja Bjelica and Jadranka Rebeka Anic (eds), *Women's Religious Voices: Migration, Culture and (Eco) Peacebuilding*, Zürich: LIT Verlag, pp. 79–93.

Barbaro, Nicole and Scott M. Pickett, 2016, 'Mindfully green: examining the effect of connectedness to nature on the relationship between mindfulness and engagement in pro-environmental behavior', *Personality and Individual Differences* 93, pp. 137–42.

Barr, Jessica Marion, 2017, 'Auguries of elegy: the art and ethics of ecological grieving', in Ashlee Cunsolo and Karen Landman (eds), *Mourning Nature: Hope at the Heart of Ecological Loss and Grief*, Montreal and Kingston; London; Chicago, IL: McGill-Queen's University Press, pp. 190–226.

Bifrost Arts Music, 2016, *Lamentations*, album recording.

Blumenthal, David, 2002, 'Liturgies of anger', *CrossCurrents* 52.2, pp. 178–99.

Bomberg, Elizabeth and Alice Hague, 2018, 'Faith-based climate action in Christian congregations: mobilisation and spiritual resources', *Local Environment* 23.5, pp. 582–96.

Boyce-Tillman, June, 2022, 'Hymns and soil theology', *Practical Theology* 15.5, pp. 1–17.

Braun, Sebastian, 2017, 'Mourning ourselves and/as our relatives: environment as kinship', in Ashlee Cunsolo and Karen Landman (eds), *Mourning Nature: Hope at the Heart of Ecological Loss and Grief*, Montreal and Kingston; London; Chicago, IL: McGill-Queen's University Press, pp. 64–91.

Brown, David, 2007, *God and Grace of Body: Sacrament in Ordinary*, Oxford: Oxford University Press.

Browning, Joseph, 2021, 'Sound and more-than-human sociality in Catherine Clover's Oh! Ah ah pree trra trra', *Organised Sound* 26.2, pp. 179–89.

Buckley, David T., 2022, 'Religious elite cues, internal division, and the impact of Pope Francis' Laudato Si'', *Politics and Religion* 15.1, pp. 1–33.

Canedo, Ken, 2017, *Laudato Si' and the Songs of our Common Home*, https://www.ocp.org/en-us/blog/entry/laudato-si-our-common-home, accessed 24.01.2024.

Capaldi, Colin A. et al., 2015, 'Flourishing in nature: a review of the benefits of connecting with nature and its application as a wellbeing intervention', *International Journal of Wellbeing* 5.4, pp. 1–16.

Cardiphonia Music, 2020, *Daughter Zion's Woe*, album recording.

CCLI, n.d.[a], *The Footprints Where Your People Tread*, https://songselect.ccli.com/Songs/5929573/the-footprints-where-your-people-tread, accessed 24.01.2024.

CCLI, n.d.[b], *This is God's World*, https://songselect.ccli.com/Songs/1182304/this-is-gods-world, accessed 24.01.2024.

Chase-Dunn, Christopher K. and Paul Almeida, 2020, *Global Struggles and Social Change: From Prehistory to World Revolution in the Twenty-First Century*, Baltimore, MD: Johns Hopkins University Press.

Chase, Linda, 2022, 'For Our Common Home, by Linda J Chase, (Adapted from Laudato Si')', *YouTube*, 24 October, https://www.youtube.com/watch?v=gotXo7dTAf8, accessed 24.01.2024.

Christian Aid, 2020, *Song of the Prophets: A Global Theology of Climate Change*, https://web.archive.org/web/20220705204317/https://www.christianaid.org.uk/sites/default/files/2020-05/song-of-the-prophets-theology-climate-change-report-May2020.pdf, accessed 24.01.2024.

Christian Aid, 2021a, *Song of the Prophets: A Requiem for the Climate*, https://archive.org/details/songs-of-the-prophets-digital-programme-aw, accessed 24.01.2024.

Christian Aid, 2021b, 'Songs of the Prophets', *YouTube*, 9 June, https://www.youtube.com/watch?v=y6LcVKEPTOg, accessed 24.01.2024.

Christian Climate Action, n.d.[a], *Climate Emergency Christmas Carol Book*, available from https://christianclimateaction.org/2019/11/25/climate-emergency-christmas-carol-book/, accessed 24.01.2024.

Christian Climate Action, n.d.[b], *Christian Climate Action Songbook*, available from https://christianclimateaction.org/resources/other-resources/songbook/, accessed 24.01.2024.

Christian Climate Action, 2020, 'Extinction Rebellion was Live: Memorial for Life, Supporting the Church Synod to Save our Children', *Facebook*, 12 February, https://www.facebook.com/christianclimateaction/posts/1460995284057516, accessed 24.01.2024.

Climate Vigil, n.d., *Climate Vigil Songs Worship Guide*, available from https://www.climatevigil.org/album, accessed 24.01.2024.

Clingerman, Forrest, 2015, 'Theologians as interpreters – not prophets – in a changing climate', *Journal of the American Academy of Religion* 83.2, pp. 336–55.

Coat of Hopes, n.d., 'Coat of Hopes', *Coat of Hopes*, https://www.coatof hopes.uk, accessed 24.01.2024.

Coleman, David, 2021, *Friends and Troublemakers: Recycled Hymns for Urgent Use*, https://www.ecocongregationscotland.org/wp-content/uploads/2021/02/Friends-and-troublemakers.pdf, accessed 24.01.2024.

Communities of the Mystic Christ, n.d., 'Forest Church', *Communities of the Mystic Christ*, http://www.mysticchrist.co.uk/forest_church, accessed 24.01.2024.

Cunsolo, Ashlee, 2017, 'Climate change as the work of mourning', in Ashlee Cunsolo and Karen Landman (eds), *Mourning Nature: Hope at the Heart of Ecological Loss and Grief*, Montreal and Kingston; London; Chicago, IL: McGill-Queen's University Press, pp. 169–89.

Cunsolo, Ashlee and Karen Landman (eds), 2017, *Mourning Nature: Hope at the Heart of Ecological Loss and Grief*, Montreal and Kingston; London; Chicago, IL: McGill-Queen's University Press.

Dall'Oglio, Cecilia, 2020, 'Ecological initiatives of the Global Catholic Climate Movement', *Studia Ecologiae Et Bioethicae* 18.1, pp. 61–72.

Damström, Cecilia, 2019, 'Requiem for our Earth', *Cecilia Damström*, https://ceciliadamstrom.com/Requiem/, accessed 24.01.2024.

Deane-Drummond, Celia, 2017, *A Primer in Ecotheology: Theology for a Fragile Earth*, Eugene, OR: Cascade Books.

Devine, Kyle, 2019, *Decomposed: The Political Ecology of Music*, Cambridge, MA: MIT Press.

Eco Congregation Scotland, 2012, *Celebrating Creation: Ideas and Resources for Worship*, http://www.ecocongregationscotland.org/wp-content/uploads/2012/10/m2.pdf, accessed 24.01.2024.

Eve, Alison, 2015, *The Song of the Wheel*, Solihull: Ritualitas.

Feld, Steven, 2012, *Sound and Sentiment: Birds, Weeping, Poetics, and Song in Kaluli Expression*, Durham, NC: Duke University Press.

Fisk, Anna, 2017, 'Appropriating, romanticizing and reimagining: pagan engagements with indigenous animism', in Kathryn Rountree (ed.), *Cosmopolitanism, Nationalism, and Modern Paganism*, New York: Palgrave Macmillan, pp. 21–42.

Flores, Nichole M., 2018, '"Our sister, mother earth": solidarity and familial ecology in *Laudato Si*"', *Journal of Religious Ethics* 46.3, pp. 463–78.

Francis, 2015, *Encyclical Letter Laudato Si' of the Holy Father Francis on Care for our Common Home*, available from https://www.vatican.va/content/francesco/en/encyclicals/documents/papa-francesco_2015 0524_enciclica-laudato-si.html, accessed 25.01.2024.

Friberg, Anna, 2022, 'Disrupting the present and opening the future: Extinction Rebellion, Fridays for Future, and the disruptive utopian method', *Utopian Studies: The Journal of the Society for Utopian Studies* 33.1, pp. 1–17.

Frith, Simon, 1996, *Performing Rites: On the Value of Popular Music*, Oxford: Oxford University Press.

Galloway, Kate, 2019, 'Introduction: music, sound, and the aurality of the environment in the anthropocene', *Yale Journal of Music & Religion* 5.2, pp. 2–7.

Gardner, Peter, Tiago Carvalho and Maria Valenstain, 2022, 'Spreading rebellion? The rise of Extinction Rebellion chapters across the world', *Environmental Sociology* 8.4, pp. 424–35, https://www.tandfonline.com/doi/full/10.1080/23251042.2022.2094995, accessed 10.03.2024.

Gen Verde, n.d., 'Bio', *Gen Verde*, https://www.genverde.it/en/bio/, accessed 24.01.2024.

Gifford, Terry, 1999, *Pastoral*, London; New York: Routledge.

Gifford, Terry, 2017, 'The environmental humanities and the pastoral tradition', in Christopher Schliephake (ed.), *Ecocriticism, Ecology, and the Cultures of Antiquity*, Lanham, MD: Lexington Books, pp. 159–74.

Gilmurray, Jonathan, 2017, 'Ecological sound art: steps towards a new field', *Organised Sound* 22.1, pp. 32–41.

Goethe-Institut, 2021, 'Requiem: in memoriam of twelve recently extinct species', *Goethe-Institut*, https://www.goethe.de/ins/us/en/sta/bos/ver.cfm?event_id=22146611, accessed 24.01.2024.

Goodyer, Ian, 2009, *Crisis Music: The Cultural Politics of Rock Against Racism*, Manchester: Manchester University Press.

Gordon, William Van, Edo Shonin and Miles Richardson, 2018, 'Mindfulness and nature', *Mindfulness* 9, pp. 1655–8.

Grainger, Brett Malcolm, 2014, 'The Vital Landscape: Evangelical Religious Practice and the Culture of Nature in America, 1790–1870', PhD thesis, Harvard Divinity School.

Green, Duncan, 2016, *How Change Happens*, Oxford: Oxford University Press.

Grey, Carmody T. S., 2020, '"The only creature God willed for its own sake": anthropocentrism in *Laudato Si'* and *Gaudium Et Spes*', *Modern Theology* 36.4, pp. 865–83.

Guest, Mathew, 2022, 'From Protestant ethic to neoliberal logic: Evangelicals at the interface of culture and politics', in Ralph W. Hood and Sariya Cheruvallil-Contractor (eds), *Research in the Social Scientific Study of Religion*, vol. 32, Leiden: Brill, pp. 482–507.

Hague, Alice, 2018, 'Faithful Advocates: Faith Communities and Environmental Activism in Scotland', PhD thesis, University of Edinburgh.

Hague, Alice and Elizabeth Bomberg, 2023, 'Faith-based actors as climate intermediaries in Scottish climate policy', *Policy Studies* 44.5, pp. 589–607, https://www.tandfonline.com/doi/full/10.1080/01442872.2022.2137122, accessed 10.03.2024.

Hall, Alan, 2014, 'Worshipping caretakers: the creation and our stewardship of it in hymnody', *The Hymn Society of Great Britain and Ireland: Occasional Papers* 3.7.

Hartman, Laura, 2017, 'Wrestling with wickedness: a response', *Worldviews: Global Religions, Culture, and Ecology* 21.1, pp. 87–95.

Heaney, Maeve, 2013, 'Music space: living "in between" the Christian and the artistic callings', in Tom Beaudoin (ed.), *Secular Music and Sacred Theology*, Collegeville, MN: Liturgical Press, pp. 16–31.

Heintzman, Paul, 2003, 'The wilderness experience and spirituality: what recent research tells us', *Journal of Physical Education, Recreation & Dance* 74.6, pp. 27–32.

Heuvel, Steven C. van den, 2018, 'The theocentric perspective of *Laudato Si'*: a critical discussion', *Philosophia Reformata* 83.1, pp. 51–67.

Howard, Jay R. and John M. Streck, 1999, *Apostles of Rock: The Splintered World of Contemporary Christian Music*, Lexington, KY: University Press of Kentucky.

Hymns Ancient & Modern, 2019, 'Theology Slam: Hannah Malcolm on Theology and the Environment', *YouTube*, https://www.youtube.com/watch?v=GknXxsvqToU, accessed 25.01.2024.

Ingalls, Monique M., 2012, 'Singing praise in the streets: performing Canadian Christianity through public worship in Toronto's Jesus in the City parade', *Culture and Religion* 13.3, pp. 337–59.

Ingram, David, 2010, *The Jukebox in the Garden: Ecocriticism and American Popular Music Since 1960*, Amsterdam; New York: Rodopi.

Jamieson, Dale, 2015, 'Theology and politics in *Laudato Si''*, *American Journal of International Law* 109, pp. 122–6.

Kendall, David, 2016, '"All nature sings, and around me rings the music of the spheres": Christianity and the transmission of a cosmic eco-musicology', in Melissa Brotton (ed.), *Ecotheology in the Humanities: An Interdisciplinary Approach to Understanding the Divine and Nature*, Lanham, MD: Lexington Books, pp. 119–39.

Kerridge, Richard, 2014, 'Ecocritical approaches to literary form and genre', in Greg Garrard (ed.), *The Oxford Handbook of Ecocriticism*, Oxford; New York: Oxford University Press, pp. 361–75.

Kidwell, Jeremy et al., 2018, 'Christian climate care: slow change, modesty and eco-theo-citizenship', *Geo: Geography and Environment* 5.2, pp. 1–18.

Klomp, Mirella, 2021, *Playing On: Re-Staging the Passion After the Death of God*, Leiden: Brill.

Krauß, Andrea, 2011, 'Constellations: a brief introduction', *MLN* 126.3, pp. 439–45.

Lansfield, Jessica Loraine, 2015, 'Interpreting Social Engagement Strategies of the Jellyfish Project Through a Social Marketing Lens: The Power of Music and Lived Experiences', PhD thesis, University of Victoria.

Latour, Bruno, 2004, *Politics of Nature: How to Bring the Sciences into Democracy*, Cambridge, MA; London: Harvard University Press.

Laudato Si' Action Platform, n.d.[a], 'About the Platform', *Laudato Si' Action Platform*, https://laudatosiactionplatform.org/about/, accessed 25.01.2024.

Laudato Si' Action Platform, n.d.[b], 'The Laudato Si' Goals', *Laudato Si' Action Platform*, https://laudatosiactionplatform.org/laudato-si-goals/, accessed 25.01.2024.

Laudato Si' Movement, 2021, 'Global Catholic Climate Movement Announces its New Name as Part of Deeper Changes', *Laudato Si' Movement*, https://laudatosimovement.org/news/gccm-announces-its-new-name/, accessed 25.01.2024.

Laudato Si' Movement, 2022, 'Laudato Si' Festival – Music Event and Talkshow from Assisi, on Eco-Spirituality', *YouTube*, 28 May, https://www.youtube.com/watch?v=-7sGnXxHF-0, accessed 25.01.2024.

Laudato Si' Week, n.d.[a], 'Biodiversity and Art: Laudato Si' Festival Gives Glory to God for Creation', *Laudato Si' Week 2023*, https://laudatosiweek.org/2021/05/22/biodiversity-art-laudato-si-festival-glory-god-creation-en-news/, accessed 25.01.2024.

Laudato Si' Week, n.d.[b], 'What is Laudato Si' Week?', *Laudato Si' Week 2023*, https://laudatosiweek.org/what-is-laudato-si-week, accessed 25.01.2024.

Lindenbaum, John, 2013, 'The neoliberalization of contemporary Christian music's new social gospel', *Geoforum* 44, pp. 112–19.

Mall, Andrew, 2021, 'Music business, ethics, and Christian festivals: progressive Christianity at Wild Goose Festival', in Nathan Myrick and Mark Porter (eds), *Ethics and Christian Musicking*, Abingdon; New York: Routledge, pp. 105–23.

Manuel, Sandesh, n.d., 'Musik', *Sandesh Manuel*, https://www.sandeshmanuel.com/, accessed 25.01.2024.

McFarland Taylor, Sarah, 2007, *Green Sisters: A Spiritual Ecology*, Cambridge; London: Harvard University Press.

McLaren, Brian, 2011, 'Open Letter to Worship Songwriters (Updated)', *Brian D. McLaren*, 21 April, https://brianmclaren.net/open-letter-to-worship-songwriters-updated/, accessed 25.01.2024.

Menning, Nancy, 2017, 'Environmental mourning and the religious imagination', in Ashlee Cunsolo and Karen Landman (eds), *Mourning Mature: Hope at the Heart of Ecological Loss and Grief*, Montreal and Kingston; London; Chicago, IL: McGill-Queen's University Press, pp. 39–63.

Moor, Joost de et al., 2021, 'New kids on the block: taking stock of the recent cycle of climate activism', *Social Movement Studies* 20.5, pp. 619–25.

Morton, Timothy, 2007, *Ecology without Nature: Rethinking Environmental Aesthetics*, Cambridge, MA: Harvard University Press.

Music of the Plants, n.d., 'About Us', *Music of the Plants*, https://www.musicoftheplants.com/about-us/, accessed 25.01.2024.

Naor, Lia and Ofra Mayseless, 2020, 'The therapeutic value of experiencing spirituality in nature', *Spirituality in Clinical Practice* 7.2, pp. 114–33.

Nche, George, 2022, 'Five years after: an overview of the response of Catholics in Africa to the Laudato Si's call for creation care', in Ezra Chitando, Ernst M. Conradie and Susan Kilonzo (eds), *African Perspectives on Religion and Climate Change*, New York; Abingdon: Routledge, pp. 120–46.

Nita, Maria, 2016, *Praying and Campaigning with Environmental Christians: Green Religion and the Climate Movement*, New York: Palgrave Macmillan.

Operation Noah, 2014, 'Hymns on the Theme of Creation', *Operation Noah*, http://operationnoah.org/wp-content/uploads/2014/06/List_of_hymns.pdf, accessed 25.01.2024.

Parks, Josh, 2022, 'Singing with our companion species', *Reformed Journal*, 11 June, https://blog.reformedjournal.com/2022/06/11/singing-with-our-companion-species/, accessed 25.01.2024.

Patrignani, Adélaïde, 2022, 'Swiss Youth Adapt Laudato Si' to Music and Bring it to the Stage', *Vatican News*, https://www.vaticannews.va/en/church/news/2022-06/laudato-si-art-music-stage-swiss-youth-pope-francis.html, accessed 25.01.2024.

Pedelty, Mark, 2016, *A Song to Save the Salish Sea: Musical Performance as Environmental Activism*, Bloomington; Indianapolis, IN: Indiana University Press.

Pedelty, Mark et al., 2022, 'Ecomusicology: tributaries and distributaries of an integrative field', *Music Research Annual* 3, pp. 1–36.

Plumwood, Val, 1993, *Feminism and the Mastery of Nature*, London; New York: Routledge.

Plumwood, Val, 2002, *Environmental Culture: The Ecological Crisis of Reason*, London: Routledge.

Porter, Mark, forthcoming, 'The intersectional ecologies of civic musical spaces', in Steve Guthrie et al. (eds), *The Oxford Handbook of Music and Christian Theology*, Oxford: Oxford University Press.

Porter, Mark, 2017, 'Sounding back and forth: dimensions and directions of resonance in congregational musicking', *Journal of the American Academy of Religion* 8.2, pp. 446–69.

Porter, Mark, 2020, *Ecologies of Resonance in Christian Musicking*, New York: Oxford University Press.

Publicover, Jennifer L. et al., 2018, 'Music as a tool for environmental education and advocacy: artistic perspectives from musicians of the Playlist for the Planet', *Environmental Education Research* 24.7, pp. 925–36.

Rähme, Boris, 2021, 'Religion and innovation: charting the territory', in Benoît Godin, Gérald Gaglio and Dominique Vinck (eds), *Handbook on Alternative Theories of Innovation*, Cheltenham; Northampton, MA: Edward Elgar Publishing, pp. 310–33.

Riis, Ole and Linda Woodhead, 2010, *A Sociology of Religious Emotion*, Oxford; New York: Oxford University Press.

Rocke, Stephanie, 2015, 'From Mass to Politicised Concert Mass', PhD thesis, Monash University.

Rosenthal, Rob and Richard Flacks, 2011, *Playing for Change: Music and Musicians in the Service of Social Movements*, Boulder, CO: Paradigm Publishers.

Rountree, Kathryn, 2012, 'Neo-paganism, animism, and kinship with nature', *Journal of Contemporary Religion* 27.2, pp. 305–20.

Russell, Roly et al., 2013, 'Humans and nature: how knowing and experiencing nature affect well-being', *Annual Review of Environment and Resources* 38, pp. 473–502.

Ryan, Robin, 2019, 'No human ever made a cathedral such as this: scoping the ecology of the carols by candlelight effect in Australia's open-air environments', *Yale Journal of Music & Religion* 5.2, pp. 64–81.

Safran, Benjamin A., 2019, '"A gentle, angry people": music in a Quaker nonviolent direct-action campaign to power local green jobs', *Yale Journal of Music & Religion* 5.2, pp. 82–102.

Sage, Vanessa, 2009, 'Encountering the wilderness, encountering the mist: nature, romanticism, and contemporary paganism', *Anthropology of Consciousness* 20.1, pp. 27–52.

Schmidt, Leigh Eric, 1991, 'From Arbor Day to the Environmental Sabbath: nature, liturgy, and American Protestantism', *Harvard Theological Review* 84.3, pp. 299–323.

Schutte, Nicola S. and John M. Malouff, 2018, 'Mindfulness and connectedness to nature: a meta-analytic investigation', *Personality and Individual Differences* 127, pp. 10–14.

Seales, Chad E., 2019, *Religion around Bono: Evangelical Enchantment and Neoliberal Capitalism*, University Park, PA: Penn State University Press.

Snell, Tristan L. and Janette G. Simmonds, 2012, '"Being in that environment can be very therapeutic": spiritual experiences in nature', *Ecopsychology* 4.4, pp. 326–35.

Summit, Jeffrey A., 2000, *The Lord's Song in a Strange Land: Music and Identity in Contemporary Jewish Worship*, Oxford: Oxford University Press.

Taylor, Bron, 2001a, 'Earth and nature-based spirituality (part I): from deep ecology to radical environmentalism', *Religion* 31.2, pp. 175–93.

Taylor, Bron, 2001b, 'Earth and nature-based spirituality (part II): from Earth First! and bioregionalism to scientific paganism and the new age', *Religion* 31.3, pp. 225–45.

Taylor, Hollis, 2017, *Is Birdsong Music? Outback Encounters with an Australian Songbird*, Bloomington, IN: Indiana University Press.

The Porter's Gate, 2020, *Lament Songs*, album recording.

The Resound Worship Songwriting Podcast, 2022, 'Ep 97 – Isaac Wardell & Climate Vigil Songs', available from https://resoundworship.org/podcast/, accessed 25.01.2024.

Titon, Jeff Todd, 2019, 'Ecojustice, religious folklife and a sound ecology', *Yale Journal of Music & Religion* 5.2, pp. 103–16.

Vatican IHD, 2021, '#LaudatoSiWeek Laudato Si Festival "Songs for Creation"', 22 May, https://www.youtube.com/watch?v=H_PVYUa_Qhw, accessed 25.01.2024.

Waisbord, Silvio, 2020, 'Family tree of theories, methodologies, and strategies in development communication', in Jan Servaes (ed.), *Handbook of Communication for Development and Social Change*, New York: Springer, pp. 93–132.

Waldron, David and Janice Newton, 2012, 'Rethinking appropriation of the indigenous: a critique of the romanticist approach', *Nova Religio: The Journal of Alternative and Emergent Religions* 16.2, pp. 64–85.

Warren, Jeff R., 2021, '"That worship sound": ethics, things, and shimmer reverberation', in Nathan Myrick and Mark Porter (eds), *Ethics and Christian Musicking*, Abingdon: Routledge.

Warson, Gillian, n.d., 'Environment, hymns of' in *The Canterbury Dictionary of Hymnology*, Norwich: Canterbury Press.

WDR, 2022, 'Mönch, Rapper und YouTuber @PaterSandeshManuel Kölner Treff WDR', *YouTube*, 3 May, https://www.youtube.com/watch?v=fvG9zqmylgY%5C&t=22s, accessed 25.01.2024.

Wenger, Etienne, 1998, *Communities of Practice: Learning, Meaning, and Identity*, Cambridge: Cambridge University Press.

Weston, Donna, Leah Coutts and Marcus Petz, 2021, 'Music and the twenty-first century eco-warrior', *SN Social Sciences* 1.9, pp. 1–23.

Whale, Helen and Franklin Ginn, 2017, 'In the absence of sparrows', in Ashlee Cunsolo and Karen Landman (eds), *Mourning Nature: Hope at the Heart of Ecological Loss and Grief*, Montreal and Kingston; London; Chicago, IL: McGill-Queen's University Press, pp. 92–116.

Wilkins, Dominic, 2020, 'Pope Francis, *Care for Creation*, and Catholic environmental imagery', *Environmental History* 25.2, pp. 361–71.

Wilkins, Dominic, 2022, 'Catholic clerical responses to climate change and Pope Francis's *Laudato Si*", *Environment and Planning E: Nature and Space* 5.1, pp. 146–68.

Wilkinson, Katharine K., 2012, *Between God and Green: How Evangelicals are Cultivating a Middle Ground on Climate Change*, New York: Oxford University Press.

Wrenn, Rachel, 2021, 'Worshiping in anger: anger at God in psalms and liturgies', *Liturgy* 36.1, pp. 19–26.

York, Michael, 1999, 'Invented culture/invented religion: the fictional origins of contemporary paganism', *Nova Religio* 3.1, pp. 135–46.

York, Michael, 2001, 'New age commodification and appropriation of spirituality', *Journal of Contemporary Religion* 16.3, pp. 361–72.

Zwissler, Laurel, 2011, 'Second nature: contemporary pagan ritual borrowing in progressive Christian communities', *Canadian Woman Studies* 29.1/2, pp. 16–23.

Index of Names and Subjects